Lecture Notes in Computer Science 1185

Edited by G. Goos, J. Hartmanis and J. van Leeuwen

Advisory Board: W. Brauer D. Gries J. Stoer

Springer
Berlin
Heidelberg
New York
Barcelona
Budapest
Hong Kong
London
Milan
Paris
Santa Clara
Singapore
Tokyo

G. Ventre J. Domingo-Pascual
A. Danthine (Eds.)

Multimedia Telecommunications and Applications

Third International COST 237 Workshop
Barcelona, Spain, November 25-27, 1996
Proceedings

 Springer

Series Editors

Gerhard Goos, Karlsruhe University, Germany

Juris Hartmanis, Cornell University, NY, USA

Jan van Leeuwen, Utrecht University, The Netherlands

Volume Editors

Giorgio Ventre
Università di Napoli Federico II, Dipartimento di Informatica e Sistemistica
Via Claudio 21, I-80125 Napoli, Italy
E-mail: ventre@nadis.dis.unina.it

Jordi Domingo-Pascual
Universitat Politècnica de Catalunya
Departament d'Arquitectura de Computadors
Campus Nord - Mòdul D6, c. Gran Capità sn, 08071 Barcelona, Spain
E-mail: jordid@ac.upc.es

André Danthine
Université de Liège, Institut Montefiore, B28
B-4000 Liège, Belgium
E-mail: danthine@ulg.ac.be

Cataloging-in-Publication data applied for

Die Deutsche Bibliothek - CIP-Einheitsaufnahme

Multimedia telecommunications and applications : proceedings
/ Third International COST 237 Workshop, Barcelona ; Spain,
November 25 - 27, 1996. G. Ventre ... (ed.). - Berlin ;
Heidelberg ; New York ; Barcelona ; Budapest ; Hong Kong ;
London ; Milan ; Paris ; Santa Clara ; Singapore ; Tokyo :
Springer, 1996
(Lecture notes in computer science ; Vol. 1185)
ISBN 3-540-62096-6
NE: Ventre, Giorgio [Hrsg.]; International COST 237 Workshop <3,
1996, Barcelona>; GT

CR Subject Classification (1991): H.4.3, C.2, I.7.2, B.4.1, H.5.1, H.5.3

ISSN 0302-9743
ISBN 3-540-62096-6 Springer-Verlag Berlin Heidelberg New York

© Springer-Verlag Berlin Heidelberg 1996
Printed in Germany

Typesetting: Camera-ready by author
SPIN 10549967   06/3142 – 5 4 3 2 1 0   Printed on acid-free paper

Third COST 237 Workshop
Multimedia Telecommunications and Applications
Barcelona, November 25-27, 1996

Preface

During the last few years we have witnessed an impressive increase in the interest in multimedia communications and applications. After an initial research phase, pilot projects fostered by national and international research programs are now being launched all over Europe and the world involving not only a limited audience of specialists, but literally bringing multimedia to the home. Consequently, the issues related to the provision of advanced multimedia services are increasingly of interest. In parallel, the need for a timely definition of standards for the design and the deployment of new technologies in this area is encouraging the creation of consortia where manufacturers, telecommunication operators and researchers can exchange ideas and experiences so that products and services may be developed more rapidly.

Since its beginning in February 1992 as an new Action in the CEC COST Program, COST 237 aimed to be a forum for a continuous and effective exchange of ideas and experiences among academia, industry and telecom operators. Since then 14 countries from the European Union and from the rest of Europe have joined the Action with more than 30 institutions directly involved in the activities. COST 237 can be considered as a network of researchers that regularly have the opportunity to exchange the research results and experiences they have developed in the framework of their own national and international scientific cooperation. Most of this exchange is accomplished through the activities of Working Groups. Originally, two groups were created on Multimedia Teleservices (WG 1) and New Transport Services Specifications (WG 2); during the last year one new Working Group has been activated in the critical area of Security and Applications (WG 4) and one more on Communication Support for Multimedia Applications (WG 3) has replaced WG 2, closed on completion of the defined tasks. While the Working Group on Security and Applications is focused on the issues of user authentication, service securing and authentication, copyright protection, and interoperability of certifications authorities, the Working Group on Communication Support for Multimedia Applications is aimed at the study of the problems related to protocols and mechanisms for enhanced transport services, multipeer communications, resource management and Quality of Service assurance, distributed signalling and control, and heterogeneity of network architectures. Another activity worth mentioning pursued by COST 237 during the last year is surely the organisation of common events and initiatives, like the first ECMAST conference in Louvain la Neuve, with other European research frameworks such as the ACTS Program.

The 1996 COST 237 International Workshop on Multimedia Telecommunications and Applications was planned by the Management Committee of the Action with the aim of emphasising such changes in the Action structure and of further encouraging interactions of its members with other research groups active in all the areas of interest. This workshop is the third of a series of successful events held formerly in Vienna (1994) and in Copenhagen (1995) that have attracted the attention of a quite large number of researchers in the field of distributed multimedia. Protocols and mechanisms for cooperative multimedia applications are currently a prime focus for many standardisation and institutional bodies. Among these are the ATM Forum, the IETF, DAVIC, TINA, ITU, and ISO. A major goal of this third workshop was to present the latest research developments that may be of interest to these various bodies and to encourage an exchange of experiences. For these reasons, in constrast to the approach followed for the 1995 Workshop, this year an open Call for Papers was issued, that produced a total of 34 papers submitted from Europe and USA. After a very selective refereeing process, less than 50% papers of high quality were accepted for a total of 15 presentations, to achieve the goal of a compact and very interesting workshop program. We are sorry that many other interesting papers had to be rejected due to the lack of space but we hope that this selection effort will be appreciated by both the authors and the attendees.

Five technical sessions have been scheduled. The first session, on Multipeer and Group Communication, includes papers presenting an innovative routing algorithm with delay constraints, a group management system for multimedia applications, and a mechanism to improve scalability in transport level multicasting techniques. The second session is totally devoted to Quality of Service control and assurance issues. The first paper introduces a platform for service creation and renegotiation; the second paper is related to the problems of specifying QoS requirements in standard distributed programming environments; the third paper, finally, presents a scheme for the conversion of communication media to be used in mobile network platforms. The third session on Applications and Teleservices includes three experiences on the development of new applications and services: a presentation on a hypermedia system for the deployment of new services, a paper on an innovative CSCW application for the large scale distribution of conferences, and a paper on a pilot interactive application. The fourth session on Multimedia Protocols and Platforms includes a paper on the design issues for a new network protocol, a coding scheme suited for mobile multimedia communication, and a presentation on the development of a multimedia conferencing application using a signalling platform. The fifth and final session is devoted to Performance issues. The first paper included presents an innovative approach to the design of a network and adaptation layers for ATM networks. The second paper deals with the capability of a Unix SVR4 kernel to host multimedia applications. The third paper presents a technique for the evaluation of the impact of optimisation techniques on the transmission of MPEG-2 video streams.

The Third COST 237 International Workshop on Multimedia Telecommunications and Applications is organised by the CEC COST 237 Action on Multimedia Telecommunications Services and this year is hosted in the wonderful city of Barcelona by the Universitat Politècnica de Catalunya with the support of Telefónica de España and the Comissionat per a Universitats i Recerca (Generalitat de Catalunya). A number of people have to be acknowledged for the preparation and organisation of this event. First of all, André Danthine and David Hutchison for their effort in leading this Action along these years on a path that can surely be said to be very successful. I also wish to thank the members of the Steering Committee for the planning of the event, and the members of the Program Committee for their effort in inviting, refereeing, and selecting the papers with the invaluable help of all the reviewers who supported us. A sincere acknowledgement goes also to all the authors who submitted their research work to our workshop. Finally, I would like to express my deep appreciation to the Organising Committee and particularly to Jordi Domingo-Pascual for taking on the critical task of making such an event possible. To all the participants and presenters I am glad to send my best wishes for an enjoyable and fruitful workshop.

October 1996   Giorgio Ventre

Program Committee

Giorgio Ventre	*U. di Napoli "Federico II", Italy* (Chair)
Patrick Baker	*HP Labs, UK*
Torsten Braun	*IBM ENC, Germany*
Augusto Casaca	*INESC, Portugal*
Geoff Coulson	*Lancaster U., UK*
Jon Crowcroft	*UCL, UK*
André Danthine	*U. of Liège, Belgium*
Michel Diaz	*LAAS/CNRS, France*
Christophe Diot	*INRIA, France*
Wolfgang Effelsberg	*U. of Mannheim, Germany*
Serge Fdida	*MASI, France*
Domenico Ferrari	*U. Cattolica à Piacenza, Italy*
David Hutchison	*Lancaster U., UK*
Villy B. Iversen	*Technical U. of Denmark, Denmark*
Marjory Johnson	*RIACS, USA*
Helmut Leopold	*Alcatel, Austria*
Benoit Macq	*U. Catholique de Louvain, Belgium*
Jordi Domingo-Pascual	*U. Politècnica de Catalunya, Spain*
Ramon Puigjaner	*U. de les Illes Balears, Spain*
Radu Popescu-Zeletin	*GMD-Fokus, Germany*
Aruna Seneviratne	*UTS, Australia*
Otto Spaniol	*T.U. Aachen, Germany*
Jean-Bernard Stefani	*CNET, Paris, France*
Ralf Steinmetz	*T.U. Darmstadt, Germany*
Harmen van As	*Vienna Institute of Technology, Austria*
Martina Zitterbart	*T.U. Braunschweig, Germany*

Steering Committee

André Danthine	*U. of Liège, Belgium* (Chair)
Theodoros Bozios	*Intracom, Greece*
Christophe Diot	*INRIA, France*
Jordi Domingo-Pascual	*U. Politècnica de Catalunya, Spain*
Wolfgang Effelsberg	*U. of Mannheim, Germany*
Serge Fdida	*MASI, France*
Domenico Ferrari	*U. Cattolica à Piacenza, Italy*
José Guimaraes	*ADETTI, Portugal*
David Hutchison	*Lancaster U., UK*
Villy Iversen	*Technical U. of Denmark, Denmark*
Borka Jerman-Blazic	*Institute Jozef Stefan, Slovenia*
Helmut Leopold	*Alcatel, Austria*
Vassili Loumos	*NTUA, Greece*
Radu Popescu-Zeletin	*GMD-Fokus, Germany*
Sandor Stefler	*PKI, Hungary*
Giorgio Ventre	*U. of Napoli, Italy*

Organizing Committee

Jordi Domingo-Pascual	*U. Politècnica de Catalunya, Spain*
Josep Solé-Pareta	*U. Politècnica de Catalunya, Spain*
Xavier Martínez-Álvarez	*U. Politècnica de Catalunya, Spain*
Joan Vila-Sallent	*U. Politècnica de Catalunya, Spain*
Ciaran O'Colmain	*Norcontel, Ireland*

This workshop is organized with the collaboration of *the CEC COST Action on Multimedia Telecommunications Services*, the *Universitat Politècnica de Catalunya (UPC)*, *Telefónica de España* and the *Comissionat per a Universitats i Recerca* (Autonomous Government of Catalonia).

The edition of these proceedings has been possible thanks to:

Contents

Session A: Multipeer and Group Communication
Chair: Serge Fdida, Laboratoire MASI, France

A Group and Session Management System for Distributed Multimedia
Applications   1
E. Wilde, P. Freiburghaus, D. Koller, B. Plattner,
Swiss Federal Institute of Technology, Switzerland

Low-Cost ATM Multicast Routing with Constrained Delays   23
A. G. Waters, J. S. Crawford, University of Kent at Canterbury, UK

Adding Scalability to Transport Level Multicast   41
M. Hofmann, University of Karlsruhe, Germany

Session B: Quality of Service
Chair: André Danthine, University of Liège, Belgium

On Realizing a Broadband Kernel for Multimedia Networks   56
M. C. Chan, J.-F. Huard, A. A. Lazar, K.-S. Lim, Columbia University, USA

Specifying QoS for Multimedia Communications within Distributed   75
Programming Environments
D. G. Waddington, G. Coulson, D. Hutchison, Lancaster University, UK

Generic Conversion of Communication Media for Supporting Personal   104
Mobility
T. Pfeifer, R. Popescu-Zeletin, Technical University of Berlin, Germany

Session C: Applications and Teleservices
Chair: Geoff Coulson, Lancaster University, UK

A Framework for the Deployment of New Services Using Hypermedia   130
Distributed Systems
Á. Almeida, INESC, Portugal

ISABEL: A CSCW Application for the Distribution of Events   137
J. Quemada, T. de Miguel, A. Azcorra, S. Pavon, J. Salvachua, M. Petit,
D. Larrabeiti, T. Robles, G. Huecas, Universidad Politécnica de Madrid, Spain

The Bookshop Project: An Austrian Interactive Multimedia Application Case   154
Study
H. Leopold, R. Hirn, Alcatel Austria AG, Austria

Session D: Multimedia Protocols and Platforms
Chair: Helmut Leopold, Alcatel Austria AG, Austria

Issues in the Design of a New Network Protocol 169
M. Degermark, Luleå University,
S. Pink, Luleå University and Swedish Institute of Computer Science,
Sweden

Source and Channel Coding for Mobile Multimedia Communications 183
A. H. Sadka, F. Eryurtlu, A. M. Kondoz, University of Surrey, UK

Developing a Conference Application on Top of an Advanced Signalling 201
Infrastructure
R. J. Huis in 't Veld, Philips Multimedia Bussines Networks,
A.-N. Ladhani, B. van der Waaij, I. A. Widya, University of Twente,
F. Moelaert El-Hadidy, J. P. C. Verhoosel, Telematics Research Centre,
The Netherlands

Session E: Performance Studies
Chair: Christophe Diot, INRIA, France

New Network and ATM Adaptation Layers for Real-Time Multimedia 216
Applications: A Performance Study Based on Psychophysics
X. Garcia Adanez, O. Verscheure, J.-P. Hubaux,
Swiss Federal Institute of Technology, Switzerland

Multimedia Applications on a Unix SVR4 Kernel: Performance Study 232
D. Bourges Waldegg, N. Lagha, J.-P. Le Narzul, Telecom Bretagne, France

Perceptual Video Quality and Activity Metrics: Optimization of Video 249
Services Based on MPEG-2 Encoding
O. Verscheure, J.-P. Hubaux,
Swiss Federal Institute of Technology, Switzerland

Index of Authors 267

A Group and Session Management System for Distributed Multimedia Applications

Erik Wilde, Pascal Freiburghaus, Daniel Koller, Bernhard Plattner

Computer Engineering and Networks Laboratory (TIK)
Swiss Federal Institute of Technology (ETH Zürich)
CH – 8092 Zürich

Abstract. Distributed multimedia applications are very demanding with respect to support they require from the underlying group communication platform. In this paper, an approach is described which aims at providing group communication platform designers with a component which can be used for powerful group and session management functionality. This component, which can be integrated into group communication platforms, is part of a system called the group and session management system (GMS). The GMS model consists of GMS user agents, which are the components to be integrated into group communication platforms, and GMS system agents which are distributed directory agents providing the distributed database which the user agents access. Communication between these two types of agents is defined in two protocols, the GMS access protocol between user agents and system agents, and the GMS system protocol between system agents. GMS also defines a number of objects and relations which can be used to manage users, groups, and sessions on a very abstract level, thus providing both group communication platform designers and programmers of distributed multimedia application with a high-level description of group communications. This approach enables a truly integrated approach for collaborative applications, where all applications, even when using different group communication platforms, can share the same database about users, groups, and sessions. The paper also contains a short description of the ongoing implementation of GMS's components.

1 Introduction

Computer communications in the last years can, when looking from the level of a user of communication systems, be seen as focusing on two aspects, which are multimedia and multipoint communications. Both aspects can be approached in a number of different ways, ranging from very low-level, being concerned only with transport technology, to very abstract models, which take multimedia multipoint communications for granted and deal only with the design of applications and their interfaces. In general, various surveys [5, 16, 17, 19, 26] have shown that the need for communication platforms supporting multipoint communications is increasing. However, we feel that there is much more work being done in the field of multimedia than in the field of multiuser communication systems.

In this paper, we describe a system which is specifically designed to support distributed multimedia applications. Our approach is that of a component which can easily be integrated into group communication platforms to provide them with sophisticated group and session management functions. This design is motivated by the following observations:

- When considering group communications, the need for a common information base for current and potential group communication participants becomes very important. Only if the properties of a group communication are known to potential participants, it is possible for them to have enough knowledge to join the group communication. Thus, a directory for information related to group communications is necessary.
- Observations of research projects implementing distributed multimedia applications show that some of the functionality is repeatedly implemented in each application. Examples for this observation are the BERKOM Multimedia Collaboration Service (MMC) described by Altenhofen et al. [1] with its Conference Directory (CD), the CoDraft system described by Kirsche et al. [16] with its multi-party communication platform, and the Joint Viewing and Tele-Operation Service (JVTOS) described by Gutekunst et al. [7] with its Session Management Service (SMS). Consequently, it would also be useful to have a reusable software component which can be used to access the directory mentioned in the first item.

  Another aspect of this issue is that, ironically, most of today's collaborative software is not able to collaborate with other collaborative software, because in each product a different model of collaboration (such as identification of users and groups, authorization, or creation of data connections) is used. We therefore also see our work as one step towards collaborative software which not only supports collaborative users, but also makes it easily possible to collaboratively use different products.

The design of the group and session management system primarily focuses on creating a model of group and session management which can be adopted for existing as well as new group communication frameworks and which provides application programmers (ie users of the group communication frameworks) with a powerful abstraction to handle group communications. In this paper we also describe an architecture which implements the group and session management model and which has been implemented at our lab. It is currently integrated into the Multipoint Communication Framework described by Bauer et al. [3], which will then serve as the first example of a group communication framework offering the above mentioned functionality.

The paper is structured as follows. Section 2 gives a short description of selected related work. In this section we will describe research as well as standardization activities. Section 3 describes the requirements which has been used when designing our model for group and session management. Section 4 then gives a description of the model as well as the architecture which we developed to implement the model. The main components here are the data model and the two protocols which are used for data transfer within our system. In Section 5, we describe the implementation of our prototype. This section focuses on the two main building blocks of our architecture, which are two types of agents. Finally, Section 6 concludes the paper and gives some final remarks as well as some discussions of our plans for the future.

2 Related Work

In the Internet world, work is going on in the Multiparty Multimedia Session Control (MMUSIC) working group of the Internet Engineering Task Force (IETF). One result of this work is a description of the Internet multimedia conferencing architecture by Handley et al. [9], which is shown in Figure 1. The component of this architecture which is most relevant for our work is the session directory, which is based on the Session Description Protocol (SDP) v2 described by Handley and Jacobson [10]. However, since SDP can only be used for session advertisement, because it is only used for the distribution of announcements, an additional protocol is required which is used for specifically inviting users to sessions. This protocol is the Simple Conference Invitation Protocol (SCIP) and is described by Schulzrinne [22]. This splitting of session relevant information into two separate protocols has been caused by the development of the mbone (a good overview of the mbone is given by Eriksson [6]), which originally was only used for multicasting sessions with a simple session announcement protocol (SDP v1). Since a more powerful support of group communications also needs a way to identify users and groups of users, we believe that a new design than the one currently being in use for the mbone is necessary.

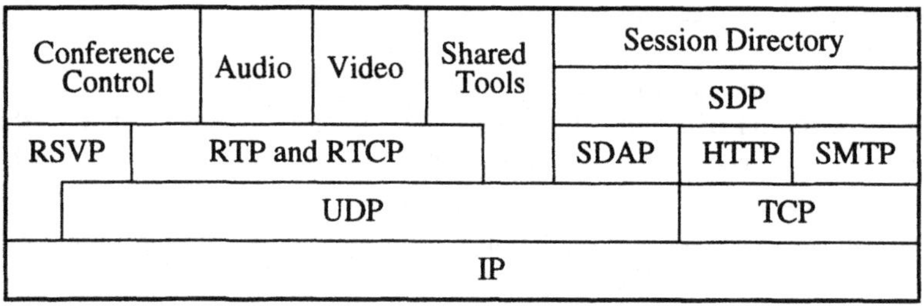

Conference Control	Audio	Video	Shared Tools	Session Directory		
				SDP		
RSVP	RTP and RTCP			SDAP	HTTP	SMTP
	UDP				TCP	
IP						

Fig. 1. Internet multimedia conferencing protocol stacks

CIO multi-peer communications as described by Henckel [11] is a very interesting concept in terms of functionality. The transport group management defined for the CIO transport service has many similarities to the model we describe in this paper. A user of the CIO multi-peer transport service uses two different components for accessing the transport service and the transport group management service. Communications are handled with two completely separate protocols. However, CIO transport group management is limited to one communications platform (ie depends on the usage of the CIO transport service), and it has not been implemented. Furthermore, since the X.500 directory service is proposed as a basis for the transport group management service, it will be impossible to have notifications sent to users, since X.500 is not capable of DSAs actively sending data to DUAs.

ITU's T.120 series of recommendations [13] is an example for a standardized architecture which also incorporates group and session management functionality. The basis of the T.120 infrastructure as shown in Figure 2 are the network specific transport protocols defined in T.123, which at the moment support data transfer using integrated services digital networks (ISDN), circuit switched digital networks (CSDN), public switched digital networks (PSDN), and public switched telephone networks (PSTN). Extensions to include future broadband networks are under study. T.123 is used by the multipoint communications service T.122/T.125, which defines a network independent service with flexible data transfer modes (broadcast and request/response), multipoint addressing (one to all, one to sub-group, and one to one), and multipoint routing (shortest paths to each receiver and uniform sequencing). Recommendation T.124 then defines a generic conference control which uses T.122's multipoint communications service. The abstract services of the conference control include create, query, join, invite, add, lock, unlock, disconnect, terminate, eject user, and transfer services. These services provide a powerful environment for implementing conferencing applications, which then use T.124 and T.122 services. However, the applicability of these recommendations is limited because only the transport infrastructures defined in T.123 can be used. The T.120 series of recommendations can therefore be regarded as one specific example which should be kept in mind when designing more general group and session management services.

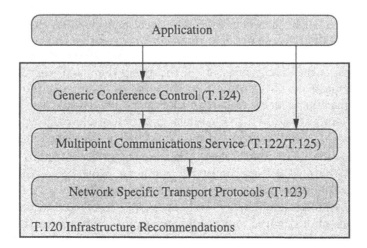

Fig. 2. Architecture of the ITU T.120 infrastructure recommendations

Other examples of multimedia communication systems dealing with group support have been described by Mauthe et al. [17]. However, all these approaches either do not concentrate on group and session management or they are restricted to certain transport infrastructures, or both. We therefore see the necessity to define a group and session management service which is independent from the transport infrastructure being used and provides a very abstract model of group communications.

3 Group and Session Management Requirements

Within this section, we identify the tasks related to group and session management which a group communication framework needs to perform. Group and session management in this context encompasses the storage and management of all information which is related to users (being the individuals working with applications), groups (which are sets of users and/or other groups), and sessions. Sessions are the representation of actual collaborations (instantiated by data exchange between distributedly working users) and can consist of one or more data flows (eg audio and video).

The main goal of the group and session management system (GMS) project is to design and develop a component which fills the gap between multipoint transport infrastructures as they are designed in actual research projects, and programmers of distributed multimedia applications. Another goal is to develop the infrastructure which is used by such a component. We therefore aim at providing these programmers with an API which is more powerful than the interfaces of common multipoint transport infrastructures. Our approach is to create a component which can be included into group communication frameworks in order to enhance their functionality. Figure 3 shows how this approach may be used, where GUA stands for GMS user agent and identifies the component provided by GMS. This component interacts with security and resource management components inside the group communication framework. It can be seen that the component is only one part of the group communication framework. Other components may be used for security purposes (in order to provide secure data communications), or for resource management, which could also include the mapping of application-level QoS parameters to network-level QoS parameters. However, we will now concentrate on the issue of group and session management.

What are the requirements such a component needs to fulfill? First and foremost, the data model used must be general enough to cover the majority of possible applications of the group communication framework. This means that both the objects and the available operations must be designed to allow a broad range of distributed multimedia applications. The abstractions needed must encompass users, groups of users, data connections between user groups, and a grouping mechanism for these data connections, because many applications need multiple data streams and it would be helpful to be able to handle these

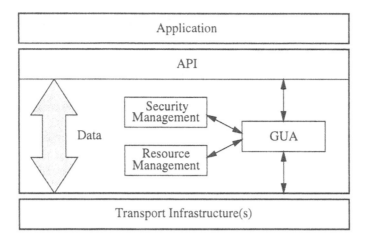

Fig. 3. Group and session management inside a group communication framework

together, especially with respect to authorization[1] and admission control[2] issues.

Furthermore, the component must be easy to integrate into existing group communication frameworks, and it must also be easy to exchange the transport infrastructure the component is using. This is a requirement because we do not want to restrict our model of group and session management to a single transport infrastructure. It has to be possible to incorporate the group and session management component into different group communication frameworks and to still be able to share as much information as possible. Naturally, transport specific information (such as addresses or QoS parameters) can not be used by group communication frameworks based on different transport infrastructures, but a lot of information, such as user identities, user authentication information, groups, and access rights, is useful independently of the transport infrastructure being used.

So far, we have mainly discussed the topics which are relevant for the support of collaborative (or distributed) applications. The aspect of multimedia communications requires additional support. However, in can be seen as a special case of communications, where data is being transmitted which requires handling with certain properties such as delay and delay jitter. In order to support distributed multimedia applications, it must be possible to specify the properties of a data connection in a way which is appropriate for multimedia data. The main idea in this area is the utilization of Quality-of-Service (QoS) parameters which make it possible to specify properties of a data connection. Because mul-

---

[1] Authorization control is performed to check whether a user is allowed to participate in a certain group communication. It is based on proper authentication of users and access privileges being assigned to objects.

[2] Admission control is used to check whether it is possible for a user to participate in a certain group communication. Admission control typically fails if there are not enough resources, such as local processing power or network capacity.

timedia data connections may have interdependencies, it must also possible to specify these. Examples for these interdependencies are synchronized data connections, which are common when individually transmitting audio and video data for video-phone applications. Another example would be the dependency of data connections when using hierarchical encodings.

As a conclusion, for the support of group and session management tasks of distributed multimedia applications, we need a system which provides connectivity throughout the lifetime of data connections, which may actively send notifications to applications, and an appropriate data model for the data inside the system. In the next section, we will discuss the model and an architecture for such a system.

4 A Model for Group and Session Management

From the requirements listed in the last section, several conclusions can be drawn. One is that a permanent database is required, which can store information about entities which are not permanently active in the context of a group communication framework, ie users and user groups. Another requirement is that certain events regarding a user of the group and session management system must be communicated to the user, and this must be initiated by the group and session management system. Therefore, it is appropriate to model the connection between the user and the group and session management system as a permanent connection during the lifetime of a user's work with a group communication framework. However, the data path of the user's application is entirely independent from the group and session management system, which distributes data with its own mechanisms. The resulting architecture of such a model is depicted in Figure 4, with GMS being the group and session management system.

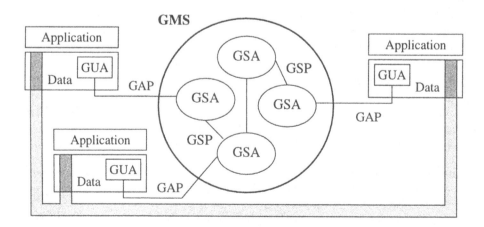

Fig. 4. GMS architecture

In this figure, GAP is the GMS access protocol, which is used as an entry point to GMS. GSP is the GMS system protocol which defines how data is exchanged inside the GMS. GAP is described in more detail in Section 4.1, GSP in Section 4.2. The main architectural components of GMS are GMS user agents (GUA) and GMS system agents (GSA), which will be described in Sections 5.1 and 5.2 respectively. While GUAs are included in group communication frameworks using GMS, GSAs are stand-alone components which, in their entirety, make up the GMS database. They are organized into domains, which are hierarchically ordered. The GMS architecture looks very similar to well-known directory services such as the Internet Domain Name System (DNS) [18] or ITU's X.500 directory service [14], but there are some notable differences.

DNS is based on the assumption that a simple lookup is sufficient (which is true for the purpose of DNS entries), while GMS requires a permanent connection between a user and the system because of the notifications which have to be delivered to GMS users. Furthermore, DNS is (for the normal user) read-only and anonymous, while GMS entries need to be modifiable and users normally are identified (and authenticated, if necessary). Therefore, the main difference between DNS and GMS is that DNS lookups are very short connections, while GMS GAP connections exist over the lifetime of a data connection, and that the typical GMS user has an identity which can be used for additional functionality, while a DNS user has anonymous access to DNS data.

The X.500 model is closer to the requirements of GMS than DNS. In fact, we first considered to use X.500 as a base for GMS, which then would have consisted simply of a set of X.500 object definitions and a software component granting access to these objects. However, there are some drawbacks with X.500, the two main disadvantages being the non-existence of DSA-initiated operations, which are needed for the notifications which are part of the GMS model, and the slow propagation of information inside X.500. Furthermore, X.500 requires the OSI stack of communication protocols, which does not fit our requirement to be able to use the access protocol over different transport infrastructures. We therefore decided not to use X.500 but to define a distributed directory service which exactly fits the needs of the GMS model. The main difference between X.500 and GMS is that GSAs are able to become active in GAP connections (instead of the purely reactive DSAs), and that GSP is designed for faster propagation of information than DSP. It is also easily possible to use GAP over a variety of transport protocols, provided they offer a reliable, connection-oriented service.

The most important definitions for GMS are the data types which are used inside the system. Because the design goal was to create a versatile model which can be used be a wide variety of distributed applications, the following object types has been defined. A complete definition of the object types can be found in the specification of GAP [23].

- *User.* A user is a person or entity using GMS. Each user has an identity (a name) and one or more ways of authenticating himself. This authentication may vary from no authentication at all (ie it is sufficient to use the right name) to sophisticated, hardware-oriented authentication schemes with multiple challenge iterations. A user object contains information about a user, such as his real world name, a description, his email address, and a list of the bindings of a user, ie the list of active GMS connections a user has.

- *Group.* GMS groups may consist of users and/or groups, depending on the definition of the group. Joining and leaving a group depends on the group's join policy and authentication requirements. Joins and leaves may be notified to a group's managers and/or members. Each group object may contain a group's real world name (eg the name of a company or a company's department), a description of the group, a group's mail address, and the access rights, which determine who is authorized to modify the attributes of the group.

- *Flow Template.* For several applications and communication platforms it is useful to have a number of predefined possibilities to set up connections. Flow templates contain information about data types which may be carried by a flow of that type, the necessary transport service, data which is needed to set up a flow of that type, information about uni- or bidirectional services, and a set of QoS parameters, which can be used to give a description of the flow template. However, flows may also be created without using a flow template.

- *Flow.* A flow is one connection for data transport. Depending on the flow's definition, it is either uni- or bidirectional, has a limited number of senders and/or receivers, and a renegotiation policy, which determines who is authorized to initiate QoS renegotiations for that flow. Flows are created when a session is created and are deleted when a session is deleted. Joining a flow takes place when a session is joined, and a flow is left when the session of a flow is left.

- *Session.* The main metaphor for group communication is a session. Each session is used to logically group a number of flows and to create an abstraction for management, authorization, and admission control for flows. The flows of a session are created when the session is created and deleted when the session is deleted. When joining a session, not all flows of the sessions must be joined, so users can choose which flows to use. Sessions may have application specific information, which consists of an application identification and application specific data, which may be interpreted by the application. Furthermore, the duration of a session may be given with either start or end times or both. In addition, it is possible to specify which authentication level a user must have to successfully join a session (provided he is authorized sufficiently). Authorization is based on the session's join policy which may be open (everyone may join), group (only members of the group associated with the session may join), or managed, which may be either relative (a given percentage of managers must confirm) or absolute (a given number of managers must confirm).

- *Certificate.* Applications with special security requirements may have the need to store certificates inside the GMS, which are used for checking data identity and integrity. Certificates include the type (which may be a predefined type or any other type), the name type (which also has a number of predefined values and the possibility to define own types), the certificate's validity, a simple name, and data and signatures, which contain the informations which is necessary for checking the data.

In the context of distributed multimedia applications, another important issue is the one of Quality-of-Service (QoS) Parameters. GMS has a very general concept of QoS specification and usage, which ranges from no QoS (which would be used for TCP/IP) up to a arbitrary number of QoS parameters which can also be renegotiated. All participants of a flow are notified of a renegotiation, so that is possible for them to take appropriate actions, such as changing the protocol parameters to adapt to the new QoS values.

GMS has four QoS parameter types, which are *unsorted values*, *sorted values*, *integer values*, and *real values*. Unsorted Values are a set of predefined values, which are not in any particular order (an example for this is a QoS parameter which defines a coding algorithm, where it is not possible to arrange the different algorithms in any order). Sorted Values are also predefined values, but these values are given as a sequence, because it is possible to arrange them in an order (an example for this type of QoS parameter is the selection of a error detection algorithm, which may be given as a sequence of none, CRC8, CRC16, and some more sophisticated algorithms). Integer and real values represent the two basic types of numbers which may be used (which may be used for throughput or error probability numbers).

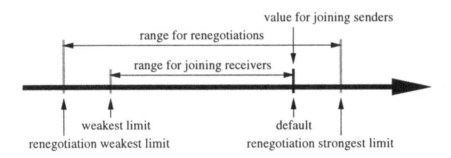

Fig. 5. Values and ranges for QoS parameters

Each flow's QoS parameters are defined by their name and type and at least a default value (which is the value used for joining participants if no local modifications are requested). Optionally, a weakest limit for joining the flow and strongest and weakest limits for renegotiations may be defined. For unordered values QoS parameters, there is no order of the values, therefore the join and renegotiation values must be given as sets of values. The concept of QoS parameter values and ranges is shown in Figure 5. When joining a session, the QoS parameters being used are taken from the flows' QoS definitions. According to the user's requirements, a weaker value than the default may be selected, as long as it does not fall short of the weakest limit for joining the flow. However, this is only possible for receivers, senders must always join with the QoS value given as the default value.

In the following sections, we will describe the two protocols of GMS, namely GAP and GSP, as shown in Figure 4. These protocols, together with the object types described earlier in this section, form the main part of the GMS model.

4.1 GMS Access Protocol (GAP)

The GMS access protocol (GAP) is used by GMS user agents to connect to the GMS. While a large part of the protocol deals with operations initiated by the GUA, there are also some messages which are initiated by the GSA the GUA is connected to. Basically, GAP requires a reliable, connection-oriented transport infrastructure. Ideally, this transport infrastructure would also allow for secure communications, but is is up to the GMS user to decide whether he requires a secure channel or not. However, the security mechanisms of GMS may be compromised if no secure GAP connection is being used. The prototype implementation of GAP is based on TCP/IP. However, this requires the group communication framework (the GUA is part of) to use TCP/IP, which could be a limitation.

If a new transport protocol for GAP should be used, both the GUA's network adaptation layer (as described in Section 5.1) and the GSA's GAP server component (as described in Section 5.2) had to be extended to support the new protocol. This way, one could easily think of a multi-protocol GSA, which for example would allow TCP/IP based as well as ISO/TP4 based GUAs to connect to it. The only limitation is that a reliable, connection-oriented protocol is required. However, it should always be kept in mind that data connections and GMS (ie GAP) connections are completely independent (as shown in Figure 4), so it is not always necessary to adapt GAP to a new transport protocol.

GAP itself can be split into three phases, each of them dealing with different aspects of the GUA-GSA interaction. A detailed specification of GAP can be found in [23]. The three phases of GAP are as follows.

– *GUA binding phase.* This is the initial phase of GAP, which is entered directly after a GUA has connected to a GSA. In this phase, the GUA has to bind to the GMS, ie it has to register with the GSA. It sends GAP version information and gets as a reply the maximum number of users supported by the GSA[3]. Furthermore, the domain name of the GSA is also sent to the GUA.

– *User authentication phase.* After the GUA has bound to the GMS, it enters the GUA bound state and is now ready to bind users. User authentication in GAP ranges from none to complex authentication schemes which require

---

[3] It is possible that such a maximum does not exist when binding to the GMS (eg because the maximum is evaluated dynamically), in this case no such number is given.

multiple data exchanged between the GSA and the GUA (ie the user)[4]. One common way of authentication is the Unix password scheme which authenticates a user by simply checking a password. However, as a result of the user authentication phase, the user is either successfully authenticated and thus bound to the GMS, or the binding attempt failed.

– *User bound phase.* Only a user who previously bound to the GMS can use the majority[5] of GAP commands. These are commands to directly access the GMS objects (such as create, modify, or delete an object), or commands which implicitly modify objects and relations, such as join and leave for groups and sessions. The authorization to perform these commands is determined by certain policies defined within the objects and the identity of the user[6]. Furthermore, in this phase it is possible that notifications about certain events are sent to the GUA.

It is important to notice that there are two possibilities to bind multiple users to the GMS. The first possibility is that every user uses his own GUA, which then opens a GAP connection to the GSA. However, this approach may fail if only one group communication framework is running on a machine which is used by several users. In this case, it is possible that several users bind to the GMS using the same GUA. GAP has been designed in a way that multiple users do not interfere when using the same GAP connection. However, because of this design, it is possible that one user is in the user authentication phase while another user is in the user bound phase. For this reason, GAP is specified using parallel state machines. A more detailed description of this design can be found in the GAP specification [23].

4.2 GMS System Protocol (GSP)

The GMS system protocol (GSP) is used by GMS system agents to communicate. This communication is necessary to exchange data and to exchange information about the configuration of GMS. A detailed description of GSP can be found in the GSP specification [24].

One of the main points about GSP is that it is a multicast-based protocol. We use the reliable, FIFO ordered (according to the definitions given by Hadzilacos and Toueg [8]) multicast protocol described by Bauer and Stiller [2] as base for GSP. The usage of a multicast based protocol is very efficient because GSAs are

---

[4] One example for a class of authentication schemes which require this type of data transfer are challenge-response schemes, which depend on a sequence of challenges sent by the server to which the client has to respond. Only if all challenges are answered correctly, the authentication is successful.

[5] Exceptions are the commands of the two other phases, which form a small subset of GAP commands.

[6] Because users may be bound to the GMS using more or less strong authentication mechanisms, GMS includes the concept of *authentication requirements*, which define for each object an authentication level by which a user must at least be bound to perform any operation regarding this object

grouped into hierarchically ordered domains, and all GSAs of a domain can be reached with a single multicast address. This way, we can use true multicasting as opposed to the multicasting mode of the X.500 directory as described in X.518 [15]. We use multicast as the method for distributing requests to all GSAs of a domain. However, when replying to a request, the replying GSA uses unicast, thus only sending the reply to the originator of the request. This approach minimizes the network load caused by the interacting GSAs.

One concept used for GSP is the one of tokens. Tokens exist inside a domain and are used to determine which GSA inside a domain in authorized to perform certain operations. The three token types of GSP are as follows, where each token exists for every domain inside the GMS.

- *Propagation of requests.* Because it is often necessary to propagate domain name resolution requests (which are described in detail later in this section) either up or down the domain hierarchy, there has to be one GSA inside each domain which is responsible for this task. Whether this role is fixed or moved from one GSA of a domain to another using some kind of load balancing strategy, is outside the scope of the GSP specification.
- *Object creation.* Object creation is also handled by multicast requests. Because only one GSA is allowed to actually create a new object when requested (otherwise duplicates would be created), the task of object creation also depends on a token. This token may be rotated among a domain's GSAs using a strategy which takes into account the storage place available on each GSA.
- *Forwarding and processing of queries.* Queries are the most processing intensive operations inside the GMS because it is necessary to search for objects matching a given pattern. Queries must either be forwarded or processed inside a domain by a dedicated GSA, which collects the results and sends them back to the originator of the query. This role is also represented by a token and may also be assigned dynamically according to some strategy.

Because tokens may get lost (eg when the machine of the token holder crashes) or may be duplicated (eg if the network has been temporarily partitioned), it is always possible for a GSA to request a token renegotiation for a domain. For this task, each GSA implements a simple finite state machine which has the three states *monitoring*, *competing*, and *tokenholding*. *Monitoring* and *tokenholding* are two stable states, indicating a GSA which does not have a token respectively does have a token. When a token renegotiation request is sent to the domain, all GSAs enter the *competing* state. By replying with claim token messages, all GSAs try to get the token. Based on the content of the claim token messages, each GSA can decide which GSA will become the token holder. This GSA then enters the *tokenholding* state, while all other GSAs will become *monitoring*. The token renegotiation process is initiated by a GSA requesting a service from a domain's GSAs and either getting no reply (ie the token was lost) or more than one reply (ie the token was duplicated) after a predefined period of time.

The basic process of operations being carried out within GSP can be separated into two phases. The first phase is the domain name resolution. As mentioned before, domains are hierarchically ordered. The GSAs of each domain only know the address of their directly superior domain and the addresses of all directly inferior domains. There is no such thing as the top-level domain but a set of top-level domains, where each top-level domain knows the addresses of all other top-level domains and the addresses of all directly inferior domains. Whenever a GSA wants to send a request to another domain, it first checks whether it has cached the address after a previous request. If not, the domain name resolution phase is started. Depending on whether the required domain is hierarchically above the requesting GSA or not (which can be decided based on the domain's name), the domain name resolution request is either sent to the superior domain or to the appropriate inferior domain. In this domain, the GSA holding the *propagation of requests* token will forward the domain name resolution request to the next domain and reply with a domain name resolution pending message to the requesting GSA. If this pending message (or more than one) is not received after a certain timeout, the requesting GSA will send a token init request to the domain, initiating a token renegotiation for this domain (as discussed in the previous paragraph). This process, which is shown in Figure 6, continues until the address of the requested domain is found, which is then directly sent back to the initiating GSA.

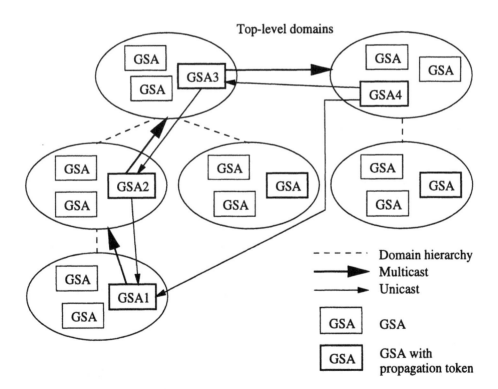

Fig. 6. GSP domain name resolution

In this figure, GSA1 is the GSA initiating the domain name resolution. GSA2, GSA3, and GSA4 each reply with a domain name resolution pending message to the GSA which sent the request to the domain using its multicast address. GSA4, which finally knows the address of the domain to which GSA1 wants to send a request, directly responds to GSA1 with the domain's address. If such a reply is not received after a timeout, the domain name resolution process is initiated again. Both this timeout and the timeout which is used to send the token init request must be chosen carefully to find the optimal balance between unnecessary repetitions respectively token renegotiations and too long idle periods. However, since we assume that the domain hierarchy will be relatively flat (eg as deep as the DNS domain name space, with a similar organization), there are not too many GSAs involved in the domain name resolution process.

Once a GSA has resolved the address to which a request must be sent (either by the process described above or from a cache, which contains addresses accessed before), the operation itself can be carried out. This is done by sending the request containing the operation to the domain. Depending on the operation, the GSAs of the requested domain behave differently. For example, when sending a modify request, the GSA storing the object will be the only one responding to the request, while all other GSAs of the domain silently ignore the request. This is possible because it is clear that only the GSA storing the object in its local database can process the request. On the other hand, if a create request is processed, first an object present request is sent to all GSAs of the domain. All GSAs of the domain send a reply, indicating whether they hold the *object creation* token or not[7]. This is necessary because the requesting GSA must be able to detect whether there is exactly one token holder. If all GSAs[8] reply with an indication that they do not hold the token (or if more than one token holder is answering), it can be concluded that a new token holder must be found and the GSA sends a token renegotiation request to the domain. Otherwise, the create request is directly sent (using unicast) to the GSA which indicated that it has the *object creation* token.

Additional protocol mechanisms exist for adding GSAs to and removing GSAs from a domain. Normally, it is assumed that a GSA is a permanently running process, but even then occasional interrupts (eg if the machine a GSA runs on is stopped) can occur. If a GSA is started, it must first join the multicast group which is assigned to the domain it is joining. Then the GSA has to send a join domain request to this address, and all GSAs of the domain reply to this request. The first reply to that request, sent by any GSA inside the domain, concludes the start up procedure and the GSA becomes part of the domain. The remaining join domain replies can be ignored. Each GSA has a local table of

---

[7] This procedure is feasible under the assumption that the number of GSAs per domain is moderate. Again, we consider an architecture similar to that of DNS, where typically each domain is served by very few servers. If the number of GSAs would be much greater than 10, the approach taken would have been prohibitive.

[8] The fact that all GSAs have replied can be concluded from the total number of GSAs in the domain, which is contained in every reply sent by a GSA.

which contain a list of triples containing an unique GSA identification, a flag indicating whether it is currently active, and a version number. This table is updated if a join domain request is received. This table is sent periodically to the domain, so that inconsistencies finally disappear. If a GSA wants to leave a domain, it sends a leave domain request to the domain, waits for the first leave domain reply, and then leaves the multicast group. The remaining GSAs update their internal tables according to the leave domain request. These tables are also updated if differences are detected between the periodically received version and the local version.

5 Implementation

In the previous section, we described the model and an architecture for the group and session management system GMS. In this section, we will briefly describe the implementation of our GMS prototype. At the time of writing, the GUA implementation has finished and the GSA implementation is still under way. As implementation base we use Sun workstations running Solaris 2.5, the Sun version of Unix. As tools we used StateMate, a commercial software for designing, simulating, and generating code for finite state machines, and snacc by Sample and Neufeld [20, 21], a public domain software for generating code for ASN.1 coding and decoding. ASN.1 coding and decoding software was required because we use ASN.1 [12] as notation language for the syntax in our protocol specifications.

5.1 GUA

The implementation of the GMS user agent, which is, together with a more detailed description of the concepts, described in [25], focuses on two aspects, the first being the easy adaptability of the GUA code to another transport infrastructure. We therefore used only abstract procedures for accessing the transport infrastructure and created a network adaptation layer which can easily be modified as long as the new transport infrastructure being used provides reliable, connection-oriented communications. This design can be seen in Figure 7.

The other important aspect is the one of control flows. Because normally the GUA can get input from both the user (at the programming interface layer) and the network (at the network adaptation layer), and we did not want to delay processing of one direction until the other direction has been handled, we use a multi-threaded design, where one thread is responsible for processing requests from the programming interface layer, while the other thread processes requests coming from the network adaptation layer. The reason why we decided to use this solution is because, as mentioned in Section 4.1, it is possible that multiple users use one GUA, and it is important to make sure that no user is able to disturb other users' work with the GMS.

The implementation of the GUA component consists of seven main building blocks. Network adaptation and programming interface are cleanly separated

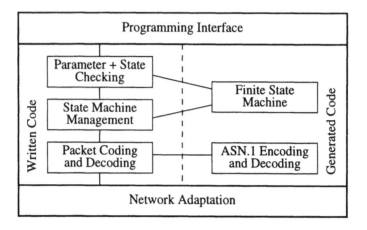

Fig. 7. GMS user agent (GUA) design

to make the code easily adaptable to different environments. Furthermore, we use generated code for the GUAs finite state machines and the encoding and decoding of GAP PDUs. This code is controlled by three components, which are responsible for checking parameters and the state of the GUA's state machine, the state machine management (ie performing state transitions and executing the appropriate actions), and the marshaling and unmarshaling of arguments.

5.2 GSA

The GSA implementation consists of three main building blocks, which are the implementation of the two protocols GAP and GSP, and the internal logic, which is responsible for performing the actions which are requested by either a GUA or a GSA. It is also possible to have GSAs which do not provide GAP access. In this case, the building block containing the GAP functionality is omitted. This architecture is depicted in Figure 8, which goes into a little more detail. In this figure, each labeled block represents a process, while the MQ components are Unix message queues, providing a convenient interprocess communication mechanism.

The main component of the GSA is the GSA manager which is the process started first during initialization. If the GSA supports GAP, the GSA manager starts the GAP server, which is then able to accept connect requests from GUAs. The GAP server currently uses TCP/IP, but it could be easily extended to support other transport protocols. The GSA manager also starts multicast processes for sending to and receiving from multicast groups. These processes implement the reliable multicast protocol mentioned in Section 4.2. Furthermore, the GSA manager starts a unicast receiver process which handles all incoming requests for the GSA. These are replies to multicast requests. The multicast modules are implemented on top of UDP/IP multicast, while the unicast receiver is based on

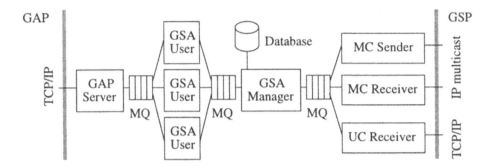

Fig. 8. GMS system agent (GSA) design

TCP/IP[9]. Finally, the GSA manager is the process which has direct access to the local database.

The GAP server handles the *GUA binding phase* described in Section 4.1. If a GUA initiates the *user authentication phase*, the GAP server dynamically starts a GSA user process which is then responsible for handling this users authentication and, if the authentication has been successful, also the *user bound phase*. The GSA user process terminates if the user unbinds, if the user's authentication failed or if the GUA forces an unbind (ie the GAP connection is terminating). Consequently, for each user bound to the GSA, a GSA user process exists which handles the users requests and delivers any results or notifications to the user.

The GSA manager has the role of the central entity in the GSA design. It accepts all incoming GSP PDUs and processes them according to their content. If a PDU is the result of a operation requested by a GSA user process, it is forwarded to this process. If an incoming GSP PDU requires the GSA to perform some actions, it either performs the appropriate actions itself or it starts a separate process (not shown in the figure) which lives as long as the operation is being processed. The GSA manager also has direct access to the database, where all local objects and relations are stored. Currently we use the standard Unix ndbm package[10] for managing the database. This package implements a simple storage for key/data pairs, in our case the pairs consist of an object's name as the key and the object's ASN.1 representation as data.

---

[9] Because we use the unicast connection only for sending a reply to the requesting GSA, a transactional variant of TCP such as T/TCP described by Braden [4] would be preferable. However, at the moment we use standard TCP, thus including the overhead of the rather expensive TCP connection establishment and the problem with both ends going to the TIME-WAIT state after closing the connection.

[10] The ndbm format is the new format for Unix databases which replaces the older dbm format and library.

5.3 Implementation results

The implementation described in the previous sections has been tested in various ways. The majority of test has been performed as load testing, where a huge load of requests was produced and the behavior of the system has been observed. After these tests (which included a number of test domains with GSAs in the local network and internationally distributed), all timeouts have been adjusted carefully to find the optimal balance between unnecessary repetitions and unnecessary wait periods. Most operations now have five or ten seconds timeouts, assuming that due to the usage of a reliable multicast protocol (which already has internal timers for keep-alive packets), timeouts on GSP level should occur only rarely.

Analyses of the GSA code showed that about 70% of the time spent is used for program logic, 30% is used for database accesses, and only 0.3% are used for coding and decoding ASN.1 data. This was a surprise to us, since the code generated by the Snacc ASN.1 to C/C++ compiler is huge (120000 lines of generated C++ code as opposed to 20000 written lines of C++ code for the various GSA processes). However, this still causes problems, because although the CPU load caused by a GSA running on a system is moderate, due to the size of the processes (most of them including coding/decoding routines), a system running a GSA is heavily loaded by swapping processes from and to memory.

One important point when discussing the performance of a distributed directory service is the scalability of the architecture. GMS can be scaled in three dimensions, which are discussed in the following list.

- *Number of users per GSA*. The current GSA implementation is obviously not suited to support a larger number of users, since every user is represented by a process (as shown in figure 8). However, the resources required for each user could be reduced to a few table entries if the GSA code was designed appropriately. Thus, the number of users per GSA could be fairly big (in the magnitude of a few hundreds) if the GSA implementation was carefully designed.
- *Number of GSAs per domain*. The number of GSAs per domain also is the number of GSAs receiving all multicast requests to this domain and replying to them, if necessary. Therefore, the number of GSAs per domain should be kept fairly small (in the magnitude of ten) to avoid the well-known implosion problem. This imposes no problem, since GSAs are meant to be central services which are remotely accessed using the GMS access protocol (GAP).
- *Number of domains*. The number of domains can be scaled in two ways. Extending the domain hierarchy horizontally does not cause any change in GSP performance, since domain name resolution is not affected by the number of subdomains of a domain. Extending the domain hierarchy vertically (ie introducing new levels of subdomains) influences the domain name resolution linearly, since the domain name resolution requests have to be propagated through more domains. Requests to domains are not influenced at all, since they are directly addressed to the domain. Furthermore, when using caching

in the GSA instead of performing a domain name resolution for every request, the effect of extending the domain hierarchy vertically could be minimized. Hence, the number of domains does not influence GMS in a way which could cause performance problems.

Consequently, GMS is able to be used in a large scale, provided the number of GSAs per domain is kept reasonably small (which is also preferable from a management point of view). We therefore believe that the approach to group and session management presented in this paper not only makes group communication platforms more flexible, but also can be used in a global scale.

6 Conclusions

In this paper we describe a group and session management system (GMS) for distributed multimedia applications. The GMS model assumes that a special component, a GMS user agent (GUA), is included into group communication frameworks which want to incorporate GMS functionality. This component then becomes an integral part of the group communication framework, ie it is not possible for application programmers to access the GUA directly. This approach has been chosen because a number of operations (especially the join session operation, which joins the requester to a number of data flows) can not be completely processed inside the GUA, but also need the group communication framework (eg for performing an admission control which needs to check whether the local and network resources are sufficient to join a session). The exchange of user data and of GMS access protocol (GAP) data is performed independently, so it is possible to use different transport infrastructures for data exchange and GAP connectivity.

Furthermore, the GMS data model is designed in a way that it allows the modeling of users, groups, and sessions in an abstract way which is suitable for different group communication frameworks. It is therefore possible to easily integrate GMS functionality into group communication frameworks, the two main issues being the abstract data model of GMS, and the separation of data and control information (GAP). The abstract data model can be used to map a framework's internal model of connections and connection handling to the abstractions used by GMS, which is then accessible to all GMS users, even if they are using different group communication frameworks. The separation of data and control information provides the framework designers with the opportunity to separate data exchange and management information, which can even be transferred using different transport infrastructures.

In addition to the standard GMS usage, where a GUA is integrated into a group communication framework, we are also implementing an application which only uses the GUA for communications and consequently can not be used for data transfer. This application could serve the same purposes than the Internet's mbone session directory (as implemented by the sd/sdr tools), but with a richer functionality (such as authentication, authorization, and the ability to use more than one group communications framework). At the time of writing, this application is in the implementation phase.

The GMS architecture is that of a distributed directory service. The distributed components are GMS system agents, communicating via the GMS system protocol (GSP). GSAs are grouped into hierarchically organized domains, which reflect organizational structures in the real world. GSP is a multicast protocol, with the unit of addressing being the domain. The protocol design is based on the assumption that domain are relatively small with respect to the number of GSAs (not much more than 10 GSAs in one domain), and that the hierarchy is relatively flat (not much more than 5 levels). Based on observations of the current structure of DNS and X.500, we believe that these assumptions are realistic. We will start an experimental GMS as soon as the GSA implementation is finished, which will give us the opportunity to adjust the timeout values (which are crucial for the proper operation of GSP) and to evaluate an internationally distributed version of GMS.

The authors would like to thank Daniel Bauer and Gerhard Nigg for providing us with the software for the reliable multicast protocol GSP is based on. We also would like to thank Murali Nanduri who implemented most of the GUA software.

References

1. Michael Altenhofen, Jürgen Dittrich, Rainer Hammerschmidt, Thomas Käppner, Carsten Kruschel, Ansgar Kückes, and Thomas Steinig. The BERKOM Multimedia Collaboration Service. In *Proceedings of ACM Multimedia 93*, pages 457–463, Anaheim, California, 1993. ACM Press.
2. Daniel Bauer and Burkhard Stiller. An Error-Control Scheme for a Multicast Protocol Based on Round-Trip Time Calculations. In *Proceedings of the 21st Conference on Local Computer Networks*, Minneapolis, October 1996.
3. Daniel Bauer, Erik Wilde, and Bernhard Plattner. Design Considerations for a Multicast Communication Framework. In *Proceedings of the Tenth Annual Workshop on Computer Communications*, Eastsound, Washington, September 1995.
4. R. Braden. T/TCP – TCP Extensions for Transactions Functional Specification. Internet RFC 1644, July 1994.
5. C. A. Ellis, S. J. Gibbs, and G. L. Rein. Groupware – Some Issues and Experiences. *Communications of the ACM*, 34(1):38–58, 1991.
6. Hans Eriksson. MBone: The Multicast Backbone. *Communications of the ACM*, 37(8):54–60, 1994.
7. Thomas Gutekunst, Thomas Schmidt, Günter Schulze, Jean Schweitzer, and Michael Weber. A Distributed Multimedia Joint Viewing and Tele-Operation Service for Heterogeneous Workstation Environments. In Wolfgang Effelsberg and Kurt Rothermel, editors, *GI/ITG Arbeitstreffen Verteilte Multimedia-Systeme*, number 5 in Praxis, Information und Kommunikation, pages 145–159, Stuttgart, Germany, February 1993. K. G. Saur.
8. Vassos Hadzilacos and Sam Toueg. Fault-Tolerant Broadcasts and Related Problems. In Sape Mullender, editor, *Distributed Systems*, chapter 5, pages 97–145. ACM Press, New York, second edition, 1993.
9. M. Handley, J. Crowcroft, and C. Bormann. The Internet Multimedia Conferencing Architecture. Internet Draft, MMUSIC Working Group, February 1996.

10. Mark Handley and Van Jacobson. SDP: Session Description Protocol. Internet Draft, MMUSIC Working Group, November 1995.
11. Lutz Henckel. Multipeer Connection-mode Transport Service Definition based on the Group Communication Framework. Technical report, GMD FOKUS, Berlin, June 1994.
12. International Organization for Standardization. Information processing systems – Open Systems Interconnection (OSI) – Specification of Abstract Syntax Notation One (ASN.1). ISO/IS 8824, 1990.
13. International Telecommunication Union. Data Protocols for Multimedia Conferencing. Draft Recommendation T.120, 1995.
14. International Telecommunication Union. The Directory – Overview of Concepts, Models and Services. Recommendation X.500, March 1995.
15. International Telecommunication Union. The Directory – Procedures for distributed operations. Recommendation X.518, March 1995.
16. T. Kirsche, R. Lenz, H. Lührsen, K. Meyer-Wegener, H. Wedekind, M. Bever, U. Schäffer, and C. Schottmüller. Communication support for cooperative work. *Computer Communications*, 16(9):594–602, 1993.
17. Andreas Mauthe, Geoff Coulson, David Hutchison, and Silvester Namuye. Group Support in Multimedia Communications Systems. In D. Hutchison, H. Christiansen, G. Coulson, and A. Danthine, editors, *Teleservices and Multimedia Communications – Proceedings of the Second COST 237 Workshop*, volume 1052 of *Lecture Notes in Computer Science*, pages 1–18, Copenhagen, Denmark, November 1995. Springer-Verlag.
18. P. Mockapetris. Domain Names – Concepts and Facilities. Internet RFC 1034, November 1987.
19. T. Rodden, J. A. Mariani, and G. Blair. Supporting Cooperative Applications. *Computer Supported Cooperative Work*, 1(1–2):41–67, 1992.
20. Michael Sample. Snacc 1.1: A High Performance ASN.1 to C/C++ Compiler. Technical report, University of British Columbia, Vancouver, July 1993.
21. Michael Sample and Gerald Neufeld. Implementing Efficient Encoders and Decoders For Network Data Representations. In *Proceedings of the IEEE INFOCOM '93 Conference on Computer Communications*, pages 1144–1153, San Francisco, 1993. IEEE Computer Society Press.
22. Henning Schulzrinne. Simple Conference Invitation Protocol. Internet Draft, MMUSIC Working Group, February 1996.
23. Erik Wilde. Specification of GMS Access Protocol (GAP) Version 1.0. Technical Report TIK-Report No. 15, Computer Engineering and Networks Laboratory, Swiss Federal Institute of Technology, Zürich, March 1996.
24. Erik Wilde. Specification of GMS System Protocol (GSP) Version 1.0. Technical Report TIK-Report No. 19, Computer Engineering and Networks Laboratory, Swiss Federal Institute of Technology, Zürich, September 1996.
25. Erik Wilde, Murali Nanduri, and Bernhard Plattner. A Transport-Independent Component for a Group and Session Management Service in Group Communications Platforms. In P. Delogne, D. Hutchison, B. Macq, and J.-J. Quisquater, editors, *Proceedings of the European Conference on Multimedia Applications, Services and Techniques*, pages 409–425, Louvain-la-Neuve, Belgium, May 1996.
26. Neil Williams and Gordon S. Blair. Distributed multimedia applications: A review. *Computer Communications*, 17(2):119–132, 1994.

Low-Cost ATM Multicast Routing with Constrained Delays

A. Gill Waters and John S. Crawford

Computing Laboratory
University of Kent at Canterbury
Canterbury, Kent, CT2 7NF
England

Abstract. An increasing number of networking applications involve multiple participants and are therefore best supported by multicasting. Where a multicast application consumes high bandwidth, it is important to minimise the effect on the network by offering economical multicast routes. Many new applications made possible by networks based on ATM involve real-time components and are therefore also delay-sensitive. This paper discusses reasonably simple techniques for multicast routing which tackle both of these constraints, that is: first, the route makes efficient use of network resources and, secondly, delays to all recipients are kept within a bound. The problem is NP-complete, so we present heuristics which build up a directed graph containing potential routing solutions and use a greedy approach to select a solution from that graph. The heuristics are discussed and evaluated and are shown to offer good results for a variety of situations including both large and small multicast groups. Our approach is also compared with a previous solution to this problem, which has a greater time complexity.

1 Introduction

Of the services envisaged for B-ISDN [13], many involve point-to-mulitpoint working. This is true not just for distribution services such as video and/or high definition TV, but also for a wide variety of interactive services including multimedia conferencing and distributed systems. These services must be properly supported at all levels in the protocol hierarchy by the network, using mechanisms which are both efficient and flexible [23].

An important support mechanism at the network layer is that of multicast routing. Assuming that the nodes are able to replicate packets or ATM cells onto appropriate outgoing links, multicast routing in integrated services broadband networks of generalised topology involves setting up a suitable tree using the nodes and links of the network. A good multicast routing strategy must consider several aspects of the service requirements and network capabilities and has two principal goals: first to offer facilities appropriate to each specific application and secondly, to make effective use of networking resources.

Multipoint services vary widely, imposing differing demands on multicast routing mechanisms. Some can be supported by a single multicast tree regardless of which participant is transmitting, easing the connection set-up procedure. Other applications, for example video distribution and multi-media conferencing will need large amounts of bandwidth and are also sensitive to delays. For multi-media conferencing, different multicast trees may be necessary for each multicast source in order to keep transmissions within acceptable delay bounds. Thus, to offer the necessary flexibility, routing algorithms must take all of these requirements into account. For the most demanding services, a compromise between efficient network use and low delay (usually an upper bound on delay) is needed. This compromise forms the basis for the heuristics discussed in the paper. Other services may be able to use simpler existing techniques; a comprehensive strategy would offer the most efficient and appropriate solutions depending on the application type.

The rest of the paper is arranged as follows. In section 2, existing techniques both for the Internet and the more general case are discussed. Section 3 presents a definition of the graph-theoretical problem which is tackled here, namely the bounded delay, minimum cost multicast tree problem. In section 4 we present two related heuristic solutions to the problem developed by the authors. We also refer to other published algorithms that address the problem, one of which we have detailed and compare with our techniques [16]. Section 5 describes the performance evaluation environment and presents results of evaluations of the heuristics which demonstrate the relative merits of the techniques. The conclusion summarises the performance of the heuristics and suggests topics for further research.

2 Approaches to Multicast Routing

Before discussing our techniques in detail, we first review existing mechanisms for multicast routing, most of which deal with specific application requirements and have a single constraint, usually either minimising cost or minimising delay. The majority of work on multicast routing has been undertaken in two different contexts. These are firstly, support for host groups on the Internet and secondly, more theoretical approaches to multicasting in networks of arbitrary topology. This section discusses both of these bodies of work.

2.1 Internet Host Groups

A host group on the Internet [6] is the collection of hosts which should receive packets sent to the group's address. The philosophy behind the use of multicasting in the Internet is an extension of the connectionless way of working on the Internet: each multicast packet is delivered to the members of the group with best efforts. The sender need not know which hosts are members of the group and indeed need not itself be a member of the group. This strategy is good for some applications, but cannot properly support all of the wide range of potential multicast applications for high-speed multiservice networks.

The initial multicast routing techniques assumed that groups are densely situated throughout the Internet and routing techniques are discussed by Deering and Cheriton in [7]. Distance vector routing (based on work by Dalal and Metcalfe in [5]) is used for multicast groups on the MBone. It builds a shortest path tree, but involves pruning and regrowing the tree to cope with potential new members. Link-state routing also provides a shortest path tree; it uses flooding to propagate membership information. Neither of these techniques scales well.

A number of recent proposals offer better scalability, the major contenders being Core Based Trees (CBT) and Protocol Independent Multicast (PIM). In both cases, receivers send specific join messages, making them more suitable for sparse groups. The protocols also address reliability, state management and evolution from existing protocols.

CBT [2] builds a single delivery tree for each host group regardless of which members may act as sources. This is done by directing information to the "core" (or one of several cores) of the group from where it is multicast to all members. CBT offers a simple and effective technique for some applications, but is not optimal where achieving low delay is important. It can also suffer from potential concentration of traffic around the core(s).

Protocol Independent Multicast (PIM) [8] defines Rendezvous Points (RPs) for the group, to which sources send messages for onward multicasting to the group. When delay is critical, it can switch from the RP tree to the shortest path from a particular source if needed.

To summarise, the Internet schemes above use one or more selections from: i) a single spanning tree which may have long delays, ii) a shortest path tree rooted at each source, iii) a common multicast tree for any source within the group, which, with appropriate choice of the centre node, has a bound of twice the maximum delay to any recipient [22]. In section 4, heuristics are introduced which offer a compromise between the first two, with a greatly improved delay bound on the third.

2.2 Theoretical Work on Multicast Routing

The general graph-theoretical problem for determining multicast routing trees is as follows. Given a connected graph $G = \langle V, E \rangle$ where V is the set of vertices and E is the set of edges, we wish to find a tree $T = \langle V_T, E_T \rangle$ where $T \subseteq G$ and T joins the vertices of a multicast group, M, where $M \subseteq V$. The tree will be selected based on some optimising criterion or criteria which depend on the cost(s) incurred in travelling along any edge in E. The graphical representation either ignores costs associated with the nodes (network switches) or includes these within the costs associated with the links and is an approach suitable for large scale networks of arbitrary topology. The majority of work has looked either at low-delay solutions or at low overall cost solutions.

Finding a multicast tree which minimises the *delay* to all of the recipients can be done quite simply using Dijkstra's algorithm. (See for example [1].) It can then be pruned of nodes and/or links which are not needed to reach the members of a multicast group. The time complexity of Dijkstra's algorithm is $O(n^2)$, where n is the number of vertices in the network.

Several algorithms are available to minimise the total *cost* of a tree which includes all the nodes in a network, e.g. Prim or Kruksal, (also described in [1]). The time complexity is similar to Dijkstra's algorithm.

When the members of a multicast group, M, are a proper subset of the set V of the vertices of the graph G, finding a minimum cost solution is harder, because it must decide whether the inclusion of vertices not in M would actually lead to a cheaper solution. The problem is well known in graph theory as the Steiner tree problem which has been shown to be NP-complete but is tractable for small networks. See for example Dreyfus and Wagner [10]. Heuristics for the Steiner tree problem have been shown to approach the ideal solution of small networks. See for example Waxman [26] or Rayward Smith [19].

Other theoretical work has concentrated on slightly different aspects. Bharath-Kumar and Jaffe consider multicast trees which trade efficiency with low average delay to recipients [3]. Kadirire introduces the concept of "geographic spread" [15] which makes it more likely that a node wishing to join an existing multicast group will find a cheap path to the tree. Jiang developed a Steiner tree variation which takes account of link capacities for high bandwidth applications, prioritising choice on high capacity links [14]).

3 Defining the Bounded Delay, Minimum Cost Multicast Routing Problem

Our concern is to find suitable solutions for a number of high speed networking applications which work within two constraints, i.e they are economical within a delay bound. The bounded delay minimum cost multicast routing problem can be stated as follows.

- Given a connected graph $G = \langle V, E \rangle$ where V is the set of its vertices and E the set of its edges, and the two functions: cost $c(i, j)$ of using edge $(i, j) \in E$ and delay $d(i, j)$ along edge $(i, j) \in E$.
- Find the tree $T = \langle V_T, E_T \rangle$, where $T \subseteq G$, joining the vertices s and $M_{k,k=1,n} \in V$ such that $\sum_{(i,j) \in E_T} c(i, j)$ is minimised and $\forall k, k = 1, n$; $D(s, M_k) \leq \Delta$, the delay bound, where $D(s, M_k) = \sum_{(i,j)} d(i, j)$ for all (i, j) on the path from s to M_k in T.

As can be seen, the general case has two cost parameters, one of which represents the delay and the other the cost of using the link. For the purpose of the following evaluations, the delays on a link will be assumed to include a fixed nominal value due to queueing in the previous node. The cost function may be proportional to real costs, it may reflect available bandwidth on the link or the total bandwidth of the link or it may relate to the link's position in the network's topology. It may in some cases be proportional to the delay parameter for the link. By separating the two parameters out, the techniques discussed here will enable network designers to incorporate their own cost strategies using these or other factors in calculating suitable multicast routes.

Note that, if the delay is unimportant, the problem reduces to the Steiner tree problem. The addition of the finite delay bound makes the problem harder, and it is still NP-complete, as any potential Steiner solution can be checked in polynomial time to see if it meets the delay bound.

4 Heuristics for Multicast Routing

For convenience we label the heuristics presented here the "Waters", "Crawford" and "Kompella et al" heuristics.

4.1 The Waters Heuristic

A simple way of finding a bounded delay multicast routing tree is to use Dijkstra's algorithm. This is the "shortest path" approach used in the Internet, but it is not optimised for the total cost of the tree. The new heuristic, the "Waters heuristic", provides an effective compromise by extending the procedure of Dijkstra's algorithm to find alternative paths which still lie within the delay bound, from which a low cost tree can be constructed.

The heuristic was first published in [24] along with some simple preliminary evaluations. This paper introduces important variations and comprehensive evaluations of these. In the first version to be introduced, the delay bound, Δ is taken to be the maximum delay from source, s, to any vertex in the network (the broadcast delay bound). Variations to this procedure will be discussed later. The procedure is as follows:

1. Use an extended form of Dijkstra's shortest path algorithm, to find for each $v \in V - \{s\}$ the minimum delay, dbv, from s to v. As the algorithm progresses keep a record of all the dbv found so far, and build a matrix $Delay$ such that $Delay(k, v)$ is the sum of the delays on edges in a path from s to v, whose final edge is (k, v). When the algorithm is complete, the maximum dbv found becomes the broadcast delay bound dbB.

2. Use dbB as the delay bound Δ. Set all elements in $Delay(k, v)$ that are greater than dbB to ∞. The matrix $Delay$ then represents the edges of a directed graph derived from G which contains all possible solutions to a multicast tree rooted at s which satisfy the delay constraint.

3. Now construct the multicast tree T. Start by setting $T = \langle \{s\}, \emptyset \rangle$.

4. Take $v \in V_T$, with the maximum dbv and join this to T. Where there is a choice of paths which still offer a solution within the delay bound, choose at each stage the cheapest edge leading to a connection to the tree.

5. Include in E_T all the edges on the path (s, v) not already in E_T and include in V_T all the nodes on the path (s, v) not already in V_T.

6. Repeat steps 4 and 5 until $V_T = V$, when the broadcast tree will have been built.

7. Prune any unnecessary branches of the tree beyond the multicast recipients.

4.2   A Worked Example

To illustrate the working of the heuristic, we take the graph shown in Fig. 1a) (also used as an example in [16]). The bracketed parameters for each link indicate (*cost, delay*). The example finds the multicast route from source F to destinations B, D, E and H.

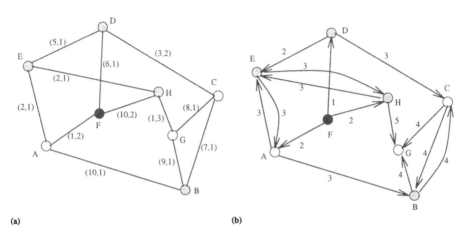

Fig. 1. Sample graph to illustrate the Waters heuristic

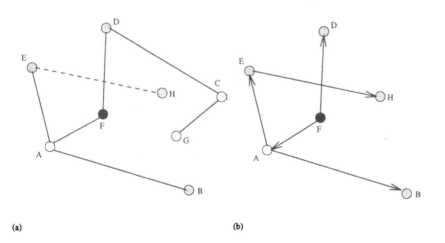

Fig. 2. Stages in construction of multicast tree

The application of the extended form of Dijkstra's algorithm results in the directed graph shown in Fig. 1b) where parameters shown against each link represent the total delay from the source, F, to reach the node at the end of that link.

The broadcast delay bound is 4 (to G). Edge HG is removed as it gives a higher delay than the bound. The tree is then constructed starting with $T = \langle F, \emptyset \rangle$. First G is connected to F using the path FD, DC, CG. B is connected via FA, AB and then E via AE, and H via EH (Fig. 2a). Finally, the edges DC and CG are pruned to give the multicast tree of Fig. 2b) whose total cost is 21 units and has a final delay bound of 4 (to H).

A minimum spanning tree, when pruned to the multicast, gives a cost of 20 units and a delay bound of 7. The tree produced by the standard use of Dijkstra's algorithm and then pruned would result in a solution with a cost of 32 units and a delay bound of 3. This simple example shows that a good compromise between delay bound and overall cost has been found by the Waters heuristic.

4.3   Time Complexity of the Heuristic

The first stage, determining the directed graph, has the same time complexity as Dijkstra's algorithm, which is at most $O(n^2)$. The vertices can be put in delay bound order during the construction of the directed graph.

In the second stage, building the multicast tree, requires a depth first search from each leaf node to find a path to the source. As the multicast tree grows the search space for each leaf to source node path becomes smaller. The time complexity of the depth first search is $O(max(N, |E|))$ [12] where N is the number of nodes, and E is the set of edges, in the leaf node to source tree. The values of N and $|E|$ depend on the topology of the network and the position of the multicast source node. Our evaluation work has shown that in practice the time required for the multicast tree construction is much closer to $O(n * m)$, where m is the maximum node degree, than $O(n^2)$ although where there are high node degree network clusters connected by very few (> 1) links, it can be much greater than $O(n^2)$.

4.4   Variations of the Waters Heuristic

A number of variations of the heuristic are possible in terms either of delay bound or choice of links to include in the multicast tree.

The tightest delay bound is the maximum of the minimum delay from source, s, to any of the destinations M_k in the multicast group. To change the heuristic to keep within this bound: first, eliminate all elements of the Delay matrix giving a greater delay and secondly, work within the multicast delay bound when constructing the tree. Without further adaptation, this method is subject to running into loops and not yielding a solution.

In setting an arbitrary delay bound, if this is between the multicast and broadcast bounds, a similar technique to the multicast bound can be used. With no bound on delay, the problem reduces to the Steiner tree problem as previously discussed. With a finite bound greater than the broadcast bound, there are two options. The broadcast bound technique described could be used, thus meeting a more stringent delay constraint than required. Alternatively, Dijkstra's algorithm could be further extended until all vertices were marked with the total delay from the source, s, found by approaching them from every neighbouring vertex. Elements of the Delay matrix exceeding the bound would be removed. This would give further alternatives for the multicast tree and could be expected to yield a lower overall cost than using a broadcast bound.

While evaluating the Waters heuristic on realistic network topologies during his MSc project, John Crawford discovered that the broadcast bound heuristic could also produce loops while examining a multicast group using the topology of Mich-Net. To combat the potential problem of looping, he put forward an alternative heuristic based on the Waters heuristic but with a different choice function. This alternative is discussed in [4]. Incidentally, at the same time Salama, Reeves, Viniotis and Sheu proposed a similar variation to the Waters heuristic in their work on multicast routing algorithms [20]. The modified version (which we call the Crawford heuristic) changes the method described in section 4.1 as follows:

- At step 1, at the same time as building the *Delay* matrix, build a corresponding matrix, *TotCost*, which gives the total cost to each vertex along the path which gives the delay in the corresponding element of the Delay matrix. i.e. If $Delay(k, j)$ represents the total delay from source s to node j, with last edge (k, j), $TotCost(k, j)$ is the total cost from the source s to node j incurred by following the same path ending in edge (k, j).
- At step 4, in constructing the path to the existing tree, instead of choosing the cheapest edge to connect to the existing tree, choose the edge which gives the cheapest total cost from the source to the vertex being considered, provided of course that the choice ensures that the delay bound is met.

An interesting and valuable property of the Crawford heuristic is that, because the total cost from the source is considered, this path will always head for the source without looping. (A loop would impose an additional and therefore greater cost.) The original Waters heuristic has now been modified to detect potential loops and remove them by using a recursive procedure whilst selecting paths back to the tree which will unwind if a loop is detected.

4.5 Heuristics with an Arbitrary Delay Bound

Several heuristics have been proposed that use arbitrary delay bounds to constrain multicast trees. The proposal of Kompella, Pasquale, and Polyzos [16], which uses a constrained application of Floyd's algorithm, is described below and used as a comparison with our heuristics. Widyono [28] proposed four heuristics based on a constrained application of the Bellman-Ford algorithm. Zhu, Parsa and Garcia-Luna-Aceves [29] based their technique on a feasible search optimisation method to find the lowest cost tree in the set of all delay bound Steiner trees for the multicast. Evaluation work carried out by Salama, Reeves, Vinitos and Sheu [21] indicate that all these heuristics have similar performance, which is generally better than the performance of our heuristics, but that the time they require to compute their solutions may be too long to be useful in large networks.

4.6 The Kompella et al Heuristic

The algorithm has three main stages.

1. A closure graph (complete graph) of the constrained cheapest paths between all pairs of members of the multicast group is found. The method to do this involves stepping through all the values of delay from 1 to Δ (assuming Δ takes an integer value) and, for each of these values, using a similar technique to Floyd's all-pairs shortest path algorithm (see [11]), which has a time complexity of $O(n^3)$. The overall time complexity of this stage is $O(\Delta n^3)$, where n is the number of vertices in the graph.
2. A constrained spanning tree in the closure graph is found using a greedy algorithm. Two alternative selection mechanisms are presented, one based solely on cost, which we used in our evaluation. The other based on cost and delay. These take $O(m^3)$ where m is the number of nodes in the multicast group.
3. The tree is regenerated in the original graph, removing any potential loops. This takes O(nm).

It should be noted that the complete graph needed for Kompella et al's solution may not exist as there will be no guarantee that the delay between any two arbitrary multicast nodes is within the delay bound. However, if these edges are set to infinity in the complete graph it will still be possible to construct a tree connecting the multicast group, provided there are paths from the source to each of the multicast nodes which fall within the delay bound.

5 Performance Modelling of the Heuristics

5.1 Network Models

Two models were used for the evaluation of the heuristics. The first is attributed to Waxman [26] and the second to Doar [9]. The Waxman model randomly distributes nodes over a rectangular coordinate grid and uses the Euclidean metric to determine the distance between them. Edges are then introduced into the network using a probability function that depends on their length.

The network model suggested by Doar is based on that of Waxman. Doar introduced a scaling factor to overcome the tendency of each node's degree to increase as the number of nodes in the network increased, a problem inherent in the Waxman model. Doar goes further by introducing hierarchical graphs as network models. These are constructed using the modified probability function to generate clusters of networks, that are then connected to a central core network using a fixed number of links. We used both network models in our evaluation work to assess the performance of our heuristics in isolated networks and networks of interconnected clusters.

We used edge lengths to represent link delay (a combination of delay attributes, e.g. propagation delay, transmission delay). Random edge costs were uniformly generated in the range [1..50] to represent the availability of resources for the edge (e.g. bandwidth).

Evaluation of the Kompella et al solution using large networks has proved impractical with our current implementation of their algorithm, because of the time required to compute the constrained cheapest paths. For these evaluations we have used the Waxman model to generate 20 random networks of 20 nodes each. The average degree of nodes in these networks is 5. Against these we find solutions for ten multicasts of each of six multicast group sizes. For the evaluation and comparison of the Waters and Crawford heuristics we have used 200 Doar hierarchical model networks each of 100 nodes. The average degree of nodes in these networks is 4 (unless stated otherwise). For each network, ten samples of each multicast group size are used.

5.2 Performance Evaluation

It is not possible to compare our solutions with optimal solutions as finding optimal solutions to the delay bound minimal cost multicast routing problem is impractical for networks other than very small ones. We compare the cost performance of the heuristics against the costs of Prim's minimum cost spanning trees [1], after pruning them to the multicast. We compare the delay performance of the heuristics against the delay bounds of Dijkstra's shortest path trees [1], after pruning them to the multicast (i.e. the multicast bound). The delay bound used for each heuristic is of course not exceeded.

Further evaluations can be found in [25]. These are carried out using the Waxman model for networks of high node degree, but the results are comparable with the more realistic cases presented here.

5.3 Comparison of Kompella et al, Waters and Crawford

Before evaluating our heuristics under a variety of conditions, we first compare them for small networks with Kompella et al's solution.

The algorithm of Kompella et al uses an arbitrary delay bound which we set equal to the broadcast delay bound, as used by our heuristics. The cost of the Kompella et al solutions will fall as the delay bound is increased because cheaper constrained cheapest paths will become available.

Figure 3 illustrates the percentage excess cost of the multicast trees found using the three heuristics under evaluation. The multicast solutions of all three heuristics outperform Dijkstra's SP solution for cost and Prim's MST for delay.

The excess costs for Dijkstra's SP solutions rise relative to Prim's MST as the multicast group size increases, since Dijkstra will add relatively fewer common edges to reach the same nodes than does Prim as the tree grows. This characteristic is common to the three heuristics because they all use shortest paths between nodes from which to select lowest cost trees, although Kompella et al use Floyd's all pairs shortest paths algorithm [1] rather than Dijkstra's SP. Interestingly, for larger groups (see figure 3) the Waters heuristic overcomes this

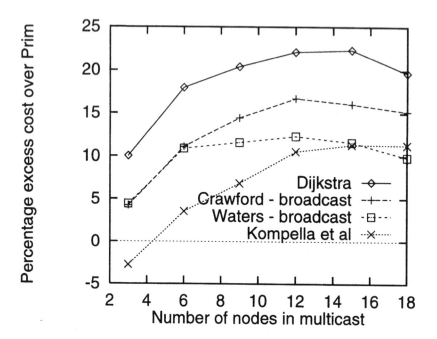

Fig. 3. Cost performance

characteristic because it has more candidate return edges to choose from when constructing return paths to the source.

The algorithm of Kompella et al generally builds multicast trees with lower costs than the solutions of Waters and Crawford, except for large multicast groups when the Waters solution becomes cheaper.

Kompella et al grows a minimum cost spanning tree from a complete graph of constrained cheapest paths between the multicast nodes by successively adding nearest cost nodes to the tree. Both the Waters and Crawford heuristics build a constrained spanning tree for the whole network, which is then pruned back to the multicast. This process in the heuristics compromises cost savings. The differences between the Waters and Crawford heuristics are examined below.

An advantage of the Kompella et al heuristic and the techniques referred to in section 4.5 is that they use an arbitrary delay bound, which can be greater than the broadcast bound or between the broadcast and multicast bounds. We are currently considering the use of arbitrary delay bounds in our heuristics (see section 4.4). The cost saving of our less time-consuming heuristics can however be seen to merit further evaluation.

5.4 Comparison of Waters and Crawford

Figure 4 illustrates the performance of the Waters and Crawford solutions for networks where edge delay and cost are independent functions.

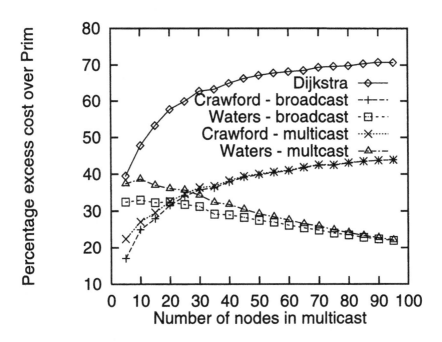

Fig. 4. Cost performance with low node degree hierarchical networks

For both the Waters and Crawford heuristics the broadcast delay bound gives cheaper solutions than does the multicast bound, because the larger bound allows a greater choice of return edges or paths than does the tighter bound.

The Crawford heuristic provides cheaper solutions for small multicast groups than does the Waters heuristic but, as the multicast group size increases relative to the network, Waters solutions become cheaper and Crawford's more expensive. The Crawford solution's excess costs increase because the heuristic mimics Dijkstra's algorithm by using costs based on the shortests paths to find return routes to the source. Its costs will generally be less than Dijkstra's because it chooses the cheapest of all the possible return paths within the delay bound, rather than the path with the least delay. The Waters heuristic constructs its low cost tree by successively adding the lowest cost edges from the highest delay nodes to build constrained paths back to the multicast source, via the existing tree, until all nodes are in the tree. As the multicast group size increases the chances of a node joining the multicast tree via its cheapest return edge will increase. As the group size increases each return path will generally have

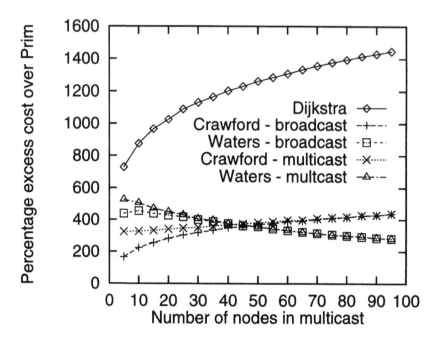

Fig. 5. Cost performance with high node degree networks

more edges, giving the algorithm more alternative paths back to the source to choose from. Both these characteristics help to significantly reduce the cost of the heuristics solutions.

The multicast group size at which the Waters heuristic begins to provide cheaper solutions than Crawford's depends on the node degree of the network. As the average node degree of the network increases the excess costs of the two algorithms begin to level out and the multicast group size at which they cross increases, as illustrated by figure 5.

When constructing the multicast trees the depth first search visited each node once only in nearly 70% of the cases for the Waters heuristic and 97.8% of the cases for the Crawford heuristic. In the worst cases the number of nodes visited by the Waters heuristic and Crawford variant were $3.44n$ and $1.71n$ respectively (where n is the number of nodes in the network). For highly connected networks (excluding dense clusters sparsely connected) the percentage of nodes visited only once during the search increased to 82.2% and 99.9% respectively.

Confidence limits at the 95% level in all graphs for mean percentage excess costs are within the range 1% to 4% for multicast group sizes in the range 20 to 95 nodes. For small multicast group sizes the variance in mean percentage excess cost increases as the proportion of mutually exclusive multicast permutations increases. Evaluations with much larger samples of smaller multicast groups give similar results within narrower confidence limits. Actual cost savings of a specific multicast depends on the topology of the network it is applied to and the position of the source node within the network.

5.5 Comparison where Edge Costs equal Edge Delays

Dijkstra gives much better cost performance when edge costs are equal to edge delays, that is where edges effectively have a single metric (figure 6). This is because, in minimising total delays to each multicast group node, it also keeps costs to each node low. Unlike Dijkstra, the Crawford solution may have a choice

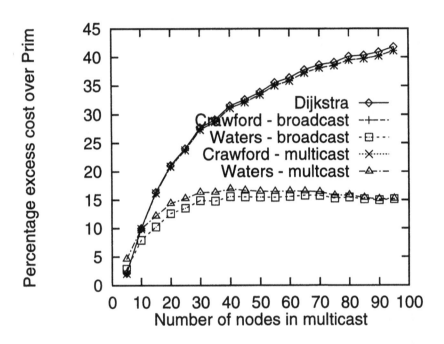

Fig. 6. Cost performance with edge costs = edges delays

of alternative return paths, albeit of the same cost, which may intersect sooner than the Dijkstra paths. It is this quality which results in Crawford's heuristic providing marginally cheaper solutions than Dijkstra's.

The Waters heuristic achieves a much lower cost than Crawford, because by choosing the cheapest constrained edges it has a greater choice of paths back to the source.

5.6 Comparison with Unit Edge Delays

In networks where delay is measured in terms of the number of "hops" between source and destination (unitary edge delay) the excess costs of all the algorithms

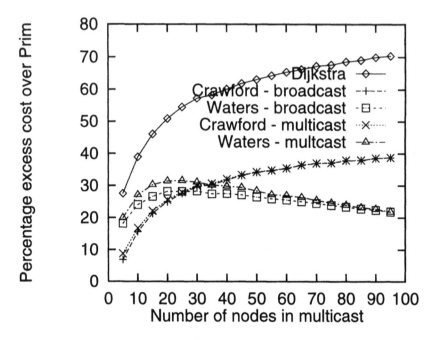

Fig. 7. Cost performance with unit edge delays

is reduced for smaller multicast group sizes (Figure 7). This is because Dijkstra's algorithm, the basis of the heuristics, will minimise the number of edges along each path when edge delay is unitary. This is not the case with non-unitary edge delays where the total path delay is minimised.

6 Scalability and Dynamicity

Scalability of multicast routing algorithms is related to both the time the algorithm requires to compute a multicast tree and the volume of state information required for the computation. Our techniques generally have low computation time, as explained in Section 4.3. The volume of state information required by a multicast routing algorithm depends on whether the multicast tree is shared or source based. Shared multicast trees are designed for use with sparse multicast groups and reduce state information requirements at the cost of increased delay. Source based multicast trees are used where delay is important or the multicast group is dense, but they use more state information. Our heuristics, being bound by an upper delay constraint are source based. A study that addresses the tradeoffs between sparse and dense mode multicasting is to be found in [27].

In practice multicast groups are often dynamic; members join and leave the multicast group throughout its lifespan thus affecting the cost and delay of the multicast tree. A simple technique for dynamic groups is to cache the multicast tree for the broadcast group. This can be used to shrink or grow the tree easily, from any group size. We are investigating even simpler techniques such as joining a new member to the tree by the least delay path and preliminary results [17] show this to be about 75% as efficient as tree recalculation, even after 2000 modifications. tree.

7   Conclusion and Further Research

Many multipoint applications envisaged for high-speed broadband networks, which include both distribution services and interactive services, will need to make efficient use of network resources while also being sensitive to delay.

We have shown that relatively inexpensive heuristics can offer solutions which balance these needs, for a variety of conditions. The performance of these heuristics compares well with the performance of more stringent, but much more costly heuristics, when the delay bound of the multicast is at the broadcast delay bound. Variations of our heuristics constrained to the multicast delay bound show how costs rise as the delay bound becomes more restrictive.

The vast majority of cases we studied completed the calculation within $O(n^2)$ time for graphs of n nodes. However, under certain conditions, the construction of leaf to source paths can lead to an unacceptably lengthy calculation due to loops or delay bound violations. A simple check for such cases would be to set a limit on the number of times nodes are considered (say $10n$) whilst constructing the tree from the undirected graph. If this limit is exceeded, the new technique could be abandoned and the tree constructed using the pruned Dijkstra's algorithm for which all information is available. This would keep calculations within $O(n^2)$ in all cases. Work is continuing to investigate the nature of the networks and multicast groups which cause this behaviour and we hope that this will lead to more efficient solutions for these special cases. (Other published techniques also include detecting and removing loops and may run into similar problems, which we have not investigated.)

In large and widely dispersed networks the broadcast delay bound may not sufficiently constrain a multipoint application. In tightly clustered networks the broadcast delay bound may be too restrictive on a multipoint application. The multicast delay bound will always be the tightest constraint on the multicast. Variations to the way the constrained delay tree is constructed in our heuristics address this issue (see section 4.4) and are currently under study.

Our heuristics are source based and assume knowledge of the network and the multicast group. The Waters heuristic performs well for larger multicast groups, whilst the Crawford heuristic is more suited to smaller multicast groups. Integration of these heuristics into a hybrid may provide better overall solutions to the delay bound minimum cost multicast routing problem. This development is also currently under study.

We believe that the performance gains of our heuristics offer a promising technique for consideration both for wide area ATM networks (and B-ISDN) and for the Internet, which is required to accommodate an increasing number of real-time multicast applications. Consequently, we are considering how our techniques can be applied to emerging multicasting standards. Preliminary results applying our work to both Core Based Trees (CBT) and Protocol Independent Multicast (PIM) have shown that they can be used to good effect [18].

8 Acknowledgements

We would like to acknowledge the support of the UK Engineering and Physical Sciences Research Council (EPSRC) for our work on multicast routing (Grant ref. GR/K55837).

References

1. A.V. Aho, J.E. Hopcroft, and J.D. Ullman. *Data structures and algorithms*. Addison Wesley, 1987.
2. A.J. Ballardie, P.F. Francis, and J. Crowcroft. Core based trees. *Computer Communications Review*, 23(4):85–95, 1993.
3. K. Bharath-Kumar and J.M. Jaffe. Routing to multiple destinations in computer networks. *IEEE Trans. on Communications*, COM-31(3):343–351, March 1983.
4. J.S. Crawford. Multicast routing: evaluation of a new heuristic. Master's thesis, University of Kent at Canterbury, 1994.
5. Y.K. Dalal and R.M. Metcalfe. Reverse path forwarding of broadcast packets. *Communictaions of the ACM*, 21(12):1040–1048, December 1978.
6. S. Deering. Host extensions for ip multicasting. rfc 1112, aug 1988.
7. S.E. Deering and D.R. Cheriton. Multicast routing in datagram internetworks and extended lans. *ACM Transactions on Computer Systems*, 8(2):85–110, May 1990.
8. S.E. Deering, D. Estrin, D. Farinacci, V. Jacobson, C-G. Liu, and L. Wei. An architecture for wide area multicast routing. *Computer Communications Review*, 24(4):126–135, October 1994.
9. J.M.S. Doar. Multicast in the Asynchronous Transfer Mode Environment. Technical Report No. 298, University of Cambridge Computing Laboratory, April 1993.
10. S.E. Dreyfus and R.A. Wagner. The steiner problem in graphs. *Networks*, 1(3):195–207, 1971.
11. R.W. Floyd. Algorithm 97: Shortest path. *Communications of the ACM*, 5(6):345, 1962.
12. Alan Gibbons. *Algorithmic Graph Theory*. Cambridge University Press, 1989.
13. ITU-T Reccomendation I.121. B-ISDN service aspects.
14. X. Jiang. Routing broadband multicast streams. *Computer Communications*, 15(1):45–51, Jan/Feb 1992.
15. J Kadirire. Minimising packet copies in multicast routing by exploiting geographic spread. *Computer Communications Review*, 24(3):47–62, July 1994.
16. V.P. Kompella, J.C. Pasquale, and G.C. Polyzos. Multicast Routing for Multimedia Communications. *IEEE/ACM Transactions on Networking*, 1(3):286–292, 1993.
17. Kuzminski.T. Alternatives for multicast routing in atm networks. Master's thesis, University of Kent at Canterbury, 1996.
18. S. Parkinson. Multicast Routing in the Internet:Evaluating Proposed Routing Mechanisms. Master's thesis, University of Kent at Canterbury, 1995.

19. V.J. Rayward-Smith. The computation of nearly minimal steiner trees. *Int. Journal of Maths, education Science and Technology*, 14(1):15–23, 1983.

20. H.F. Salama, D.S. Reeves, I. Vinitos, and Tsang-Ling. Sheu. Comparison of Multicast Routing Alogrithms for High Speed Networks. Technical report, North Carolina State University, September 1994.

21. H.F. Salama, D.S. Reeves, I. Vinitos, and Tsang-Ling. Sheu. Evaluation of Multicast Routing Algorithms for Distributed Real-Time Applications of High-Speed Networks. In *Proceedings of the 6th IFIP Conference on High-Performance Networks (HPN'95)*, 1995.

22. D. Wall. *mechanisms for broadcast and selective broadcast*. PhD thesis, Stanford University, 1980.

23. A.G. Waters. Multicast Provision for High Speed Networks. In A. Danthine and O. Spaniol, editors, *4th IFIP Conference on High Performance Networking*, pages G1.1–G1.16. Springer Verlag, December 1992.

24. A.G. Waters. A new heuristic foir atm multicast routing. In *2nd IFIP Workshop on Performance Modelling and Evaluation of ATM Networks*, pages 8/1–8/9, July 1994.

25. A.G. Waters. *Multi-party communication over packet networks*. PhD thesis, University of Essex, UK, 1996.

26. B.M. Waxman. Routing of Multipoint Connections. *IEEE journal on selected areas in communications*, 6(9):1617–1622, 1988.

27. L. Wei and D. Estrin. Multicast Routing in Dense and Sparse Modes:Simulation Study of Tradeoffs and Dynamics (95-613). Technical report, Computer Science Department, University of Southern California, 1995.

28. R. Widyono. The Design and Evaluation of Routing Algorithms for Real-time Channels. Tr-94-024, University of California at Berkeley and International Computer Science Institute, September 1994.

29. Q. Zhu, M. Parsa, and J.J. Garcia-Luna-Aceves. A Source-Based Algorithm for Near-Optimum Delay-Constrained Multicasting. In *Proceedings of INFOCOM*, pages 377–385, 1995.

Adding Scalability to Transport Level Multicast

Markus Hofmann

Institute of Telematics, University of Karlsruhe
Zirkel 2, 76128 Karlsruhe, Germany
Phone: +49 721 6083982, Fax: +49 721 388097
E-Mail: m.hofmann@ieee.org

Abstract: Driven by the capabilities of modern high speed networks, a new generation of distributed systems is emerging, that will be able to support a wide range of applications in distributed environments. Such systems are expected to consist of a large number of geographically dispersed components interacting via communication networks. Therefore, forthcoming communication systems must scale well in respect to both, number of users and number of connections. Powerful and scalable control mechanisms are essential to overcome the implosion problem and to realize efficient error recovery. This paper presents a new control scheme suitable for multicast communication in large-scale groups and describes its integration in a well-known transport protocol.

Keywords: Reliable Multicast, Implosion Control, Local Groups, Large-Scale Networks, Dynamic Groups

1. Introduction

The availability of high speed networks and recent enhancements concerning processing capacity has stirred excitement among the computerized community. Emerging networks and communication services make the support of superior distributed applications technically feasible, while increasing demand on communication in distributed environments has made it necessary. Some of these new applications, such as collaborative work, video-conferencing, or information dissemination, need to distribute user data among multiple participants. At the same time, widespread availability of IP multicast [6] and development of applications deployed in the MBone [11] have considerably increased the geographic extent and the size of communication groups. Extensive use of services like Internet radio or large-scale conferencing leads to thousands of receivers being involved in a single multicast communication. In addition, communication participants may be spread all over the world. As the size and the geographic span of communication groups increases, efficient connection management schemes including scalable error and traffic control become more and more essential. These protocol elements are placed in the transport layer of a protocol stack according to the Internet protocol architecture [3]. Recent research projects deployed new transport level protocols to meet the requirements of large-scale multicast applications in heterogeneous networks [12], [7], [10]. Some proposals only address particular aspects of group communication and focus only on some specific user environments. However, multimedia applications often have to handle several highly diverse data streams at the same time. A system supporting distributed cooperative work, for example, might offer shared use of a text editor while simultaneously providing audio and video communication between all the participants. Such applications require multipurpose and flexible communication subsystems. On the other hand, other proposals suffer from a high degree of complexity and are too general in some issues.

This paper outlines a solution to the problem of multicasting to large groups based on the *Local Group Concept (LGC)* [8]. Section 2 discusses the problems specific to the envisaged scenario and presents a brief summary of possible solutions. Section 3 and 4 introduce the Local Group Concept and a new communication service supporting the establishment and reconfiguration of local groups. Section 5 describes the integration of LGC into the Xpress Transport Protocol (XTP) [13], while Section 6 presents some results of a performance evaluation. Finally, Section 7 concludes the paper.

2. Control Schemes for Multicast Communication

Multicast capable networks provide efficient and scalable routing of data packets to multiple receivers. However, the bearer service provided by these communication networks does not fit the requirements of some applications. IP multicast, for example, offers an unreliable datagram service. The provision of full-reliable data transfer requires additional mechanisms in higher level communication protocols. Some kind of interaction between sender and receivers is necessary to ensure correct data delivery as well as to perform any kind of congestion or flow control. Neither of the hosts has enough information to control data streams on its own. The provision of full-reliable data transfer, for example, is based on a comparison between sent and received data. The transmitter has knowledge about which data units have been sent and the receivers about which data units have been received successfully. Therefore, the provision of a full-reliable communication service requires the transmission of receiver status back to the sender or vice versa.

So called sender-based schemes, in which the transmitter is responsible for controlling data transfer, rely on collecting status information at the sending site. Receivers transmit control units, including acknowledgments and traffic control information, back to the sender. As the number of receivers becomes very large, the multicast sender is overwhelmed with return messages of its receivers. This fact is known as *sender implosion*. The effect of implosion is twofold. Firstly, the large number of return messages results in processing overhead at the sender and, therefore, delays data transfer. Secondly, an enormous amount of control units may cause an excess of both, bandwidth and bufferspace, which in turn causes additional message losses. An optimal control procedure would reduce the number of status reports received by the sender down to one.

Other approaches use receiver-based schemes, where receivers themselves are responsible for error detection and error recovery. They need not to return status reports or acknowledgments to the transmitter. These schemes are not suitable to provide a full-reliable communication service, because the failure and dropping out of a single receiver could not be detected by the transmitter. Furthermore, flow or rate control always require some kind of status exchange between sender and receivers. This will lead to sender implosion even in the case of receiver-based reliable multicast communication.

Another issue of great importance is the development of efficient error correction schemes. Several procedures have been developed. Forward error correction (FEC), for example, has been proposed to be suitable for real-time applications since it allows error recovery without adding any delay associated with retransmissions. However, FEC does not prevent sender implosion caused by messages necessary for traffic control. Furthermore, adding redundant control information wastes bandwidth even when no or just a few errors occur. The sender must add enough control information to enable correction of all errors. While receivers in heterogeneous networks have highly diverse error characteristics, it is not adequate to choose a single fixed level of redundant coding. Such a fixed coding level may be excessive for some receivers while it may be insufficient for others. Other schemes use retransmissions to close the gaps in the data stream of the receives. Most of these protocols use Go-Back-N or selective repeat to retransmit lost and corrupted data. Receivers do request missed data directly from the transmitter without any consideration of network topology and current network load. In the case of group communication, it is also possible to exchange data with other receivers. It is preferable to request lost and corrupted data from a group member placed next to the host which is missing some information. An optimal error correction scheme would stimulate retransmissions of missed data units by the receiver located closest to the failing host. This would minimize transfer delay and network load.

According to the intention of the developers, several multicast protocols using different procedures have been designed. Some protocols, such as MTP (Multicast Transport Protocol) [1] and RMP (Reliable Multicast Protocol) [14] realize a centralized control scheme to provide a totally ordered multicast delivery. The central component controlling data transfer may become a bottleneck when dealing with large communication groups. Therefore, these protocols will not scale well with the number of receivers. The first protocol defining some kind of implosion control was the Xpress Transfer Protocol (XTP) [5], [13]. It defines two heuristics called *damping* and *slotting*. These algorithms suppress redundant control messages by multicasting return messages to the whole group. Recently, SRM (Scalable Reliable Multicast) [7] enhances the original XTP mechanisms and adapts them to the Internet environment. It implements receiver-based error control and is, therefore, not able to provide a full-reliable service. To avoid a flood of retransmission requests, SRM uses timers carefully set and adjusted to the current network load. However, the correct setting of timers will be no easy task for highly dynamic networks with quickly changing load and frequently changing network structure.

The implosion problem could also be solved by defining so called local groups. A new concept has been defined using a hierarchy of local groups to relieve the multicast transmitter of the responsibility for dealing with all status messages from individual receivers. This concept and some related work will be introduced in the next section.

3. Implosion Control Based on Local Groups

To avoid sender implosion and to realize a more efficient error recovery, the *Local Group Concept (LGC)* [8] splits the global communication group into separate subgroups. These subgroups will combine communication participants within a local region, forming so called *Local Groups*. Each of them is represented by a *local Group Controller* that collects status information from the members of its local group. The local Group Controller evaluates these return messages, combines them into a single control packet and transmits it back to the multicast sender. Local Group Controllers also support the provision of local retransmissions and local message processing. They are able to perform and to coordinate retransmissions of lost and corrupted data within their subgroup. This reduces delay caused by retransmissions across wide area links and decreases the load for sender and network. The integration of message processing capabilities into local Group Controllers reduces the implosion problem of multicast traffic and error control for large groups. Local Group Controllers evaluate received control units and inform the multicast sender about the status of the local group. This includes error reports as well as parameters to control data flow. Parallel processing of status reports and their combination into a single message per local group relieves the multicast sender as it reduces the number of control units to be evaluated at the sending side.

In each local region, one of the receivers is determined to function as local Group Controller. The dedicated system has to collect control messages from all the members of its subgroup and has to forward them to the multicast sender in a single composite control unit. Controllers of subgroups are also responsible for organizing local retransmissions. After evaluating received status messages a local Group Controller tries to transfer lost data units to all the receivers that have observed errors or losses. To retransmit data units a local Group Controller can use either unicast or restricted multicast transmission. This decision may be static or dynamically based on the number of failed receivers. If a controller itself misses some data units, it will try to get them from another member of its local group. Therefore, a multicast sender has only to retransmit messages missed by all members of a subgroup. Local retransmissions lead to shorter delays and decrease the number of data units flowing through a global network.

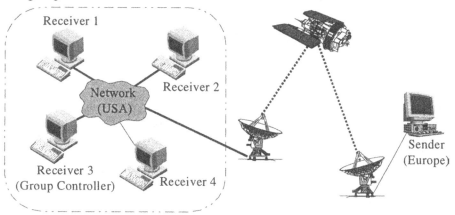

Fig. 1. Example for the Definition of Local Groups

An example scenario illustrating the basic idea and the advantage of this concept is given in Figure 1. A multicast sender located in Europe communicates over a satellite link with four receivers, which are spread over the USA. The satellite link is characterized by high transfer delay and high carrier fees. Therefore, it is desirable to reduce data traffic over this link. In this type of scenario it is useful to combine all four receivers into a single subgroup. One of the receiving hosts has to function as the controller of the subgroup. In this case, local retransmissions do not traverse the transatlantic link. This reduces transfer delay and global network load.

The resulting data streams are shown in Figure 2. The transmitter multicasts data units directly to all group members using a multicast capable delivery service (1). The local Group Controllers are kept out of the outgoing data path avoiding an extra handling of data units at each level of the hierarchy. Therefore, local Group Controllers need not to store data fragments, reassemble complete data units, interpret and forward them. Instead, multicast forwarding is done by the delivery service in a more efficient way. After receiving a status request, regular receivers transmit control messages to their corresponding local Group Controller (2). The controller combines the status reports into a single control unit and sends it to the multicast transmitter (3). Therefore, it could be said that the Local Group Concept causes some kind of triangulation of data flow.

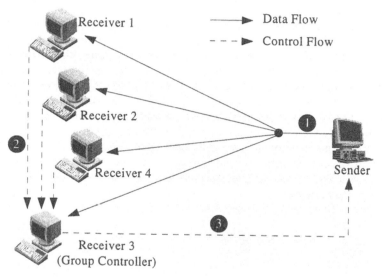

Fig. 2. Triangulation of Data Flow

The Local Group Concept also includes mechanisms to increase fault tolerance [8]. The Fail Stop of a local Group Controller, for example, is handled by dynamic reformation of Local Groups. Of course, actions performed after the failure of a communication participant depend on the given group semantic. If the communication user requires a full-reliable service, the multicast connection will be closed due to the failure of a group member. In the case of semi-reliable service, the actions to be performed depend on the attributes and the role of the failed receiver. Dynamic group reformation is also used to adapt the group structure to the current network load and to the current state of all communication participants. For further improvements in respect to transfer delay and implosion, local groups are organized in a hierarchical structure.

In the given example, the decision to combine all four receivers into one single subgroup was based on the intention of minimizing transfer delay. In other scenarios, it may be appropriate to minimize other parameters or to use a combination of several metrics. The next section discusses several aspects of structuring global communication groups.

4. GDS - A Service to Establish Local Groups

The efficiency of a concept like LGC is mainly influenced by the decision where to place the functionality of local Group Controllers. It can be integrated in several types of communication nodes, for example network switching entities, dedicated servers or regular hosts. The choice of a site for a local Group Controller will be determined by several criterion. It will be influenced by the network topology, the current network load, the capabilities of all the communication nodes and by the application specific requirements.

4.1. Placement of Local Group Controllers

The integration of controller functionality in network switching entities (e.g. switches, gateways, routers) seems to be very attractive. It allows buffering of data packets and gathering of status information along the path between transmitter and receivers. Controllers placed within network switching entities will always be located upstream in relation to the receivers. This is desirable, because downstream controllers will miss the same packets as receivers located closer to the transmitter. On the other hand, this approach will not scale well with the number of multicast connections. Additional per-connection processing and buffering is required and additional state information has to be managed at the switching entities. Controlling data flow requires examination and evaluation of individual packets. Therefore, switching entities must reassemble and evaluate packet headers and control messages before forwarding them. This increases transfer delay, especially in cell-oriented networks such as ATM. As the number of multicast connections increases, a switching entity coordinating retransmissions, processing acknowledgments and forwarding packets for several multicast connections will soon become a bottleneck. It is preferable to parallelize the handling of different multicast connections and to distribute it to several communication nodes. The CP protocol mentioned in [12], for example, uses local exchanges to combine status messages of receivers within a local region. This implies modifications of internal network switching entities. The entities have to examine and interpret the protocol control information included in data units, meaning that part of the protocol must reside on these systems. For that reason, it is difficult to update the protocol and to introduce a new protocol version.

Another solution would be the establishment of dedicated servers to coordinate multicast communication. However, this requires installation and maintenance of additional network components. The static and centralized character of this approach causes additional problems in highly dynamic networks and communication groups.

For the reasons listed above, LGC integrates the functionality of local Group Controllers into regular hosts. In principle, every communication participant is able to perform the tasks necessary for controlling a subgroup. One member per local group is designated as the active controller. The selected station represents the local group and performs local acknowledgment processing and local retransmissions. Over and above that, the flexibility of this solution permits to realize the approaches described above. If an intermediate system or a dedicated server subscribes to the multicast group, it will be able to take the role of a local Group Controller. Because the placement of local Group Controllers is dynamic in LGC, it is necessary for joining receivers to detect existing Local Groups and to chose an appropriate one. The selection of the "best-fit" local group is supported by a new service, which is introduced in the following paragraph.

4.2. Selection of Local Group Controllers

The establishment of local groups is done during connection setup. Before receiving any data, new communication participants have to subscribe to a nearby local group in respect to a chosen metric. The suitability of a certain metric, such as delay, bandwidth, throughput, error probability, reliability, carrier fees, or number of hops between two nodes, depends mainly on the application using a communication service. While an interactive application may wish to minimize transfer delay, a user transferring files is interested in reducing the financial cost of a transfer. Of course, it could also be suitable to combine several metrics and to weight them according to the intention of the service user. The establishment of local groups is also influenced by the structure and the type of a network used for data exchange. In wide-area networks or interconnected LAN/WAN environments, it is suitable to divide global communication groups into several clusters depending on the location of each host. In local environments, it is more convenient to establish logical subgroups according to the capacity of each communication participant. Inefficient receivers, for example, are associated with different subgroups to avoid asymmetrical and unfair load of Group Controllers.

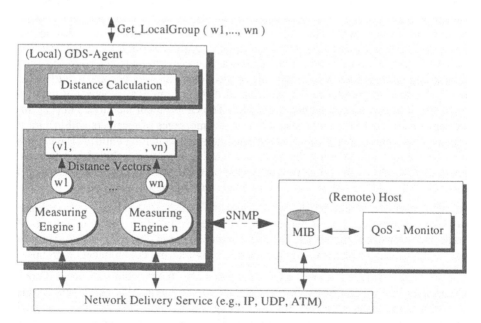

Fig. 3. GDS Architecture

The establishment of local groups is supported by a novel communication service, named *Group Distance Service (GDS)*. It provides mechanisms to find a nearby local group for each joining communication participant. Therefore, every host supporting the Local Group Concept incorporates a so called GDS Agent. As shown in Figure 3, a GDS Agent consists of several measuring engines. Each of them gauges the distance $d(i)$ to nearby located local Group Controllers based on a certain metric. One engine, for example, computes the number of hops between two hosts by implementing an expanded ring search. Another engine determines the transfer delay between two hosts based on the measured round trip time. Other engines may use SNMP (Simple Network Management Protocol) [4] to request the (financial) cost for data transfer across a certain link.

As illustrated in Figure 3, a search engine has been defined accessing the Management Information Base (MIB) of remote QoS Monitors. The Quality-of-Service (QoS) monitors store all QoS-related information (e.g. transfer delay, throughput) in this MIB [9]. A GDS agent may query the remote MIB via SNMP and use the information obtained from the remote QoS Monitor to select an appropriate local Group Controller.

Each measured value $d(i)$ is multiplied with a metric-specific weight $w(i)$ that has been defined by the service user according to its requirements. A user interested in minimizing transfer delay, for example, weights transfer delay relatively high compared to other metrics. To consider only one single metric $m(i)$, all weights are assigned a value of "0" with the exception of weight $w(i)$. The weighted values define a n-dimensional distance vector $v = (v(1), ... , v(n))$, where $v(i) = d(i) * w(i)$. The GDS Agent computes such a distance vector for each local Group Controller in its neighborhood. Therefore, it assigns each local Group Controller a member of a n-dimensional codomain. The joining receiver is assigned the zero vector. The overall distance between the new receiver and each local Group Controller is now defined to be the euclidean distance between their corresponding distance vectors.

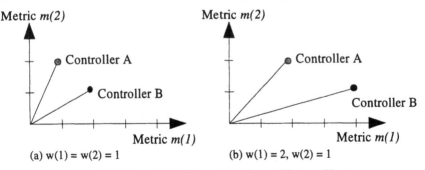

Fig.4. Impact of Different Weights on Distance Vectors

An example illustrating the dependency between the weight of metrics and the overall distance is given in Figure 4. The distance of Controller A has been computed to be 1 based on metric $m(1)$ and to be 2 in respect to metric $m(2)$. The distance of Controller B has been calculated to be (2,1).

If the user assigns the same weight $w(1) = w(2) = 1$ to both metrics, the distance of both controllers will be fixed to the same value (see Figure 4a). If the user prefers minimizing metric $m(1)$, it may assign the weights $w(1) = 2$ and $w(2) = 1$. This results in a smaller distance for Controller A, as illustrated in Figure 4b.

The GDS Agent chooses the local Group Controller "closest" to the joining receiver and returns its identity. If there is no suitable local Group Controller, the GDS user will get the information to establish a new Local Group and to appoint itself as its controller.

5. Adding Scalability to XTP Multicast

The generic character of the Local Group Concept allows an easy integration of the concept into existing multicast protocols without the necessity of substantial modifications. Therefore, LGC could be used to improve the scalability of well-known communication protocols. It is not necessary to introduce new protocols nor to establish any additional network components. This section proposes some extensions to the *Xpress Transport Protocol (XTP)* [13] to enhance its scalability based on the Local Group Concept.

XTP defines a set of protocol mechanisms whose functionality is orthogonal to one another. The intention of its design has been the provision of some basic functionality, from which the user can compose a service according to its requirements. Support for multicast communication and group management has been part of XTP since revision 3.6. First ideas for minimizing control traffic from the receivers resulted in damping and slotting algorithms to avoid implosion. Instead of returning control units exclusively to the multicast transmitter, receivers send them after a random delay to the whole group. Therefore, every group member receives this message and skips its status report if the incoming control unit corresponds to its personal state. This mechanism might reduce the number of acknowledgments to be processed by the sender. But in large-scale global networks, the usefulness of multicasting control units is very questionable. A large number of receivers together with this mechanism may cause a flood of control units all over the global network. In addition, the suppression of status messages from the receivers prevents the support of a full-reliable data transfer. In revision 4.0 [13], the mechanisms were totally revised. The suppression of status messages was removed and procedures to uniquely identify each communication participant were introduced. An addendum to the XTP 4.0 Specification [2] introduces modified multicast mechanisms to allow a more efficient mapping between control units and the corresponding communication participants. However, as the number of receivers grows larger, the sender will soon be overwhelmed with return messages of the receivers. To overcome the implosion problem and to improve the efficiency of error recovery, the Local Group Concept described in this paper has been integrated into XTP. There are some minor changes necessary, while the principle design of XTP is kept intact.

The notation used in this paper follows the definitions given in [2]. The abbreviations used are:

Kg	transmitter's local key for the multicast group
Kg'	transmitter's local key for the multicast group as a return key
Kr	receiver's local key
Kr'	receiver's local key as a return key
Ki	key assigned by the transmitter to uniquely identify receiver i
Ki'	the same as a return key
TAg(m)	multicast transport address of the group

TAt(u)	unicast transport address of the transmitter
TAr(u)	unicast transport address of the receiver
DA	destination address field in an address segment
SA	source address field in an address segment
Ag(m)	multicast delivery service address of the group (e.g. IP multicast address)
At(u)	unicast delivery service address of the transmitter (e.g. IP address)
Ar(u)	unicast delivery service address of the receiver (e.g. IP address)
dest	destination address used by the delivery service
src	source address used by the delivery service

First of all, a joining receiver has to determine an appropriate local group. It will get the identity of the local Group Controller either by using the Group Distance Service or manually from the service user. If no appropriate local group is found, the new receiver appoints itself as controller of a new group. In this case, it joins the association by multicasting a JCNTL packet according the XTP Specification. The packet sequence and the procedures will be the same as for regular receiver-initiated multicast as defined in [2]. Otherwise, the joining receiver registers itself at the chosen local Group Controller by sending a LCNTL packet to the controller. The format and the function of a LCNTL packet is very similar to the JCNTL packet defined in XTP. The difference between these two packet types enables a Group Controller to distinguish between control units used to register a receiver at the local group and a JCNTL used to register a host at the multicast transmitter. The LCNTL packet is assigned packet type 8 and it has the layout illustrated in Figure 5. It is derived from the JCNTL packet by adding a lookup segment.

Fig. 5. Format of a LCNTL Packet

As defined in XTP, mapping incoming packets onto an appropriate context requires either a return key ("abbreviated context lookup") or the regular key and the source address of the sender ("full context lookup"). Because data packets in LGC are multicasted by the transmitter, new receivers have to learn the multicast key and the address of the multicast transmitter. Local Group Controllers include this information in the LCNTL answer in response to a register request. It is stored in the Lookup Segment of the LCNTL. This segment contains all the information necessary for a receiver to associate multicast packets from the transmitter with the correct local context. The Lookup Segment format is shown in Figure 6.

Instead of defining a single lookup scheme for a certain delivery service, LGC provides parametric lookup information where one of several formats can be used. The *lkey* (lookup key) field contains the key associated with the multicast context, the transmitter's local key for the whole multicast group. Therefore, the field is 8 bytes wide. The *lformat* (lookup information format) field is 1 byte wide and identifies the format of the lookup information included in the *lookup info* field. The

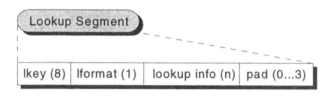

Fig. 6. Fields of the Lookup Segment

length of the lookup info field depends on the chosen lookup information format. At the moment, two formats according to Table 1 are defined.

Lformat		*Lookup Information*
Decimal	*Hex*	
0	0x00	empty
1	0x01	delivery service address of transmitter (IP Format)

Table 1 - Lookup Information Formats

When a receiver wishes to join an existing multicast group, the packet sequence is as follows. First, the receiver sends a LCNTL packet to its local Group Controller to register itself and to learn about the multicast key and the unicast address of the multicast sender. The receiver needs this information to associate multicast packets from the multicast sender with the appropriate context. The LCNTL sent by the joining receiver includes the following fields:

```
dest   =  Ac(u)       // delivery service address of controller
src    =  Ar(u)       // delivery service address of receiver
key    =  0           // defined by XTP
alloc  =  n           // current window size for the receiver
xkey   =  Kr'         // receiver's local key as a return key
DA     =  TAg(m)      // transport address of multicast group
SA     =  TAr(u)      // transport address of receiver
Tspec  =  spec(r)     // traffic specifier offered by the receiver
```

This LCNTL must have the SREQ set according to the XTP semantic. The local Group Controller checks the (local) join request and the Tspec included in the packet. If the request is denied for some reason, it will send a DIAG packet. Otherwise, it will transmit a LCNTL with the following fields:

```
dest   =  Ar(u)       // delivery service address of receiver
src    =  Ac(u)       // delivery service address of controller
key    =  Kr'         // receiver's local key as a return key
xkey   =  Ki'         // key assigned by the controller to uniquely identify
                      // the receiver (as a return key)
DA     =  TAr(u)      // transport address of receiver
SA     =  TAc(u)      // transport address of controller
Tspec  =  spec(m)     // current traffic specifier for multicast group
```

linf   = [At(u), Kg]   // lookup information necessary to associate
                                      // multicast packets with the multicast context

The lookup segment of the packet includes all the information necessary to perform a full context lookup for multicast data packets and to map them onto the appropriate context.

After the sequence has completed, the local Group Controller knows the key (Kr') to be used when sending control messages to the new receiver, and the transport level address ($TAr(u)$) for this receiver. Therefore, the local Group Controller is able to uniquely identify the new member of the group. Until now, the transmitter has no knowledge about the identity of the new receiver. It is up to the local Group Controller to ensure that all the members remain active, and that everyone gets all of the data. The local Group Controller is able to detect the failure of a receiver and it is able to ensure correct data delivery to all of its members. In the case of an error (e.g., a lost packet or a failed receiver), the local Group Controller will operate according to the given reliability class. If the reliability class of the association is "full-reliable", for example, the local Group Controller will abort the association. Otherwise, it may drop the failing host. Data flow towards single receivers can be controlled by defining "roles" or "attributes" for each receiver. The local Group Controller will control data flow according to these attributes. For example, return messages from a receiver with the attribute NOFLOW will not be used to calculate flow control parameters for the local group.

In some situations, it may be wishful to let the transmitter know the identity of all receivers. Therefore, the transmitter should be able to learn about all the group members if it so wishes. The need to identify each single receiver is indicated by the selection of a specific reliability class (RCLASS). The reliability class is part of the traffic specifier as defined in [13]. It is included in JCNTLs, LCNTLs, and TCNTLs. We have introduced an additional reliability class (RCLASS 5) indicating the need to uniquely identify all communication participants. If the transmitter has chosen this reliability class for a certain association, it will get the identity of all receivers from the local Group Controllers. An additional key exchange is not necessary, because the transmitter will never get back control messages directly from the members of a local group.

During data transfer, every multicast packet sent from the transmitter to the whole group contains the multicast key and will be addressed to the multicast group:

dest   = Ag(m)   // multicast delivery service address of the group
                                  // (e.g. IP multicast address)
src   = At(u)   // network address of transmitter
key   = Kg   // transmitter's local key for the multicast group

The transmitter cannot use a return key, because receivers will use different keys to identify their local context. Therefore, targets have always to perform a full context lookup to map the packets onto the appropriate context.

Control messages from receivers to their corresponding local Group Controller are sent using the return key assigned by the local Group Controller during the join progress:

dest	= Ac(u)	// network address of controller
src	= Ar(u)	// network address of receiver
key	= Ki'	// key assigned by the controller to uniquely identify
		// the receiver (as a return key)

Because targets use a return key, the local Group Controller can use abbreviated context lookup to map to the appropriate context.

Communication between local Group Controllers and the transmitter follows the regular XTP specification.

To request missed or corrupted data packets from members of the local group, a controller sends an ECNTL packet directly to one of the targets that have successfully received the packet. A group member receiving an ECNTL evaluates the *span* fields and transmits the requested data packets directly to the local Group Controller.

6. Performance Evaluation

Analytical methods have been applied and simulations have been performed in order to evaluate the benefits of the Local Group Concept. The simulations have been based on BONeS/ Designer, an event-driven network simulation tool by Comdisco. The simulation scenario consists of a sender and various receivers which are connected to a network (similar to the scenario in Figure 1). The sender is linked to the network across a backbone link. A transfer delay of 20 ms for the backbone link was assumed. The transfer delay within the network was assumed to be 2 ms. Both values were obtained from *ping* measurements in a real environment. The error probability was set to 10^{-1}, which is not uncommon for highly loaded internetworks. Other simulation parameters were data rate, processing delay, status request rate and burst length. Packet losses were simulated using a Markov-Model. For traffic generation, a constant data source has been used.

The evaluation examined the impact of group size on average transfer delay and network load. The values obtained for the Local Group Concept have been compared to the corresponding results for a common sender-based technique using multicast retransmission. The results show significant improvements for the new concept compared to common multicast techniques (Figure 7). More detailed information concerning the simulation scenarios, the simulation parameters and the simulation results can be found in [8].

Another interesting feature of the Local Group Concept is the partial independence between network load and number of receivers. Analytical methods have been applied to a scenario similar to that one described above. The number of receivers varied from 1 up to 1000. We assumed 50 additional receivers at the sending side of the wide area link. For the common, sender-based technique, the ratio of retransmissions and total data traffic over the wide area link depends directly on the number of recipients. The Local Group Concept, in contrast, is not influenced by the number of participants and shows very low values compared with the results of the other techniques [8].

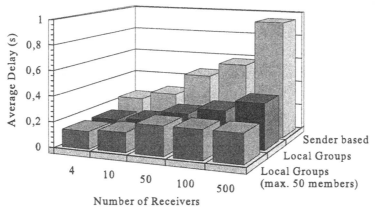

Fig. 7. Impact of Group Size on Transfer Delay

The independence between network load and the number of receivers simplifies resource reservation for highly dynamic communication groups. The bandwidth required on certain links can be reserved regardless of the current or the future group size. This simplifies the handling of multimedia data flows.

7. Conclusion

Common multicast protocols are a significant improvement over simple point-to-point protocols. However, most of them are not suitable for the case where the transmitter has to handle data flow to a large number of receivers. To avoid sender implosion and to increase efficiency of error recovery, the Local Group Concept (LGC) has been introduced. It aims for the provision of a scalable, retransmission-based error recovery in heterogeneous networks. The establishment of local groups is supported by the Group Distance Service that permits an application depended evaluation of several metrics. Performance evaluation including simulation and analysis of network load shows for the Local Group Concept significant improvements compared to common multicast control schemes. The benefits are achieved without the necessity to modify internal network equipment such as ATM switches or IP routers.

At the moment, work is going on to implement a GDS Agent on top of IP. It will be part of a complex test-suite for group communication in the Internet. The test-suite will be used to investigate the dependency between group structure (including placement of local Group Controllers) and communication performance. Furthermore, simulations of more complex scenarios will be finished and evaluated soon.

8. Acknowledgments

The author would like to thank J. William Atwood from Concordia University Montreal for valuable discussions. Special thanks to Stoyan Kenderov and other members of the LGC Working Group for various suggestions concerning the simulation models and the design of the Group Distance Service.

References

[1]   S. Armstrong, A. Freier, K. Marzullo: *Multicast Transport Protocol.* RFC 1301, February 1992.

[2]   J.W. Atwood: *Revisions to the XTP 4.0 Specification.* Message sent to the xtp-relay mailing list.

[3]   B. Carpenter: *Architectural Principles of the Internet.* RFC 1958, June 1996.

[4]   J. Case, M. Fedor, M. Schoffstall, C. Davin: *Simple Network Management Protocol (SNMP).* Network Working Group, RFC 1157, May 1990.

[5]   G. Chesson, ed.: *Xpress Transfer Protocol Specification, Revision 3.6.* Protocol Engines, Inc., Available from XTP Forum, Santa Barbara, 1993.

[6]   S. Deering, D. Cheriton: *Multicast Routing in Datagram Internetworks and Extended LANs.* ACM Transactions on Computer Systems, 8(2):85-110, May 1990.

[7]   S. Floyd, V. Jacobson, S. McCanne, C. Liu, L. Zhang: *A Reliable Multicast Framework for Light-weight Sessions and Application Level Framing.* Computer Communication Review, Vol. 25, No. 4, Proceedings of ACM SIGCOMM'95, August 1995.

[8]   M. Hofmann: *A Generic Concept for Large-Scale Multicast.* In: Broadband Communications (ed.: B. Plattner), Lecture Notes in Computer Science, No. 1044, Proceedings of 1996 International Zurich Seminar on Digital Communications, Springer Verlag, February 1996.

[9]   M. Hofmann, C. Schmidt: *A Flexible Protocol Architecture for Multimedia Communication in ATM-Based Networks.* Proceedings of 18th Biennial Symposium on Communications, Kingston, ON, Canada, June 1996.

[10]   H. Holbrook, S. Singhal, D. Cheriton: *Log-Based Receiver-Reliable Multicast for Distributed Interactive Simulation.* Computer Communication Review, Vol. 25, No. 4, Proceedings of ACM SIGCOMM'95, August 1995.

[11]   V. Kumar: *Mbone: Interactive Multimedia on the Internet.* New Riders Publishing, Indianapolis, Indiana, USA, 1995.

[12]   S. Paul, K. Sabnani, D. Kristol: *Multicast Transport Protocols for High Speed Networks.* Proceedings of International Conference on Network Protocols, Boston, 1994.

[13]   W.T. Strayer, ed.: *Xpress Transport Protocol Specification, Revision 4.0.* Available from XTP Forum, Santa Barbara, USA, March 1995.

[14]   B. Whetten. T. Montgomery, S. Kaplan: *A High Performance Totally Ordered Multicast Protocol.* Submitted to INFOCOM'95, April 1995.

On Realizing a Broadband Kernel for Multimedia Networks

Mun Choon Chan, Jean-François Huard, Aurel A. Lazar and Koon-Seng Lim

Department of Electrical Engineering
and
Center for Telecommunications Research
Columbia University, New York, NY 10027-6699

Abstract

We describe a prototype implementation of a broadband kernel, an open programming environment that facilitates the easy creation of network services and provides mechanisms for efficient resource allocation. We present a service creation methodology and show how scalable multimedia network services can be constructed from a set of broadband kernel services. As a proof of concept, we built a multi-party teleconferencing service upon the broadband kernel. The service creation framework has the following characteristics: binding of different classes of transport protocols to the application, dynamic renegotiation of application QOS using the signalling network and, support for multiple resource reservation algorithms.

1. Introduction

We describe a prototype implementation of a broadband kernel, *i.e.*, an operating platform that supports the creation of a set of services *(e.g.,* connection manager, route manager, admission controller, QOS mapper, *etc.)*. The architecture of the broadband kernel is based on the G-model of the XRM [2]. The broadband kernel incorporates the Binding Architecture [1] consisting of an organized collection of interfaces, called the *Binding Interface Base* [2] (BIB), and a set of algorithms that run on top of these. It also incorporates a native ATM stack for information transport.

The principal aim of the broadband kernel is to provide an open programming environment that facilitates the easy creation of network services and, mechanisms for efficient resource allocation. By open, we mean that the architecture must support functional Application Programming Interfaces (APIs) for developing useful services. By programmable, we mean that these APIs should be 'high-level' enough to allow the service specification and creation process to be carried out via a high-level programming language.

BIB interfaces provide an open and uniform access to abstractions that model the 'local' states of networking resources. Binding algorithms play a key role in service creation through the process of interconnecting (binding) of networking resources. QOS is explicitly modeled in the architecture via a set of abstractions that characterize the multiplexing capacity of networking and multimedia resources under QOS requirements. The resource allocation mechanisms that are integral part of the broadband kernel will not be discussed in this paper.

We present a service creation methodology and show how scalable multimedia network services can be constructed from a set of broadband kernel services. As a proof of concept, we built a multi-party teleconferencing service upon the broadband kernel. The service creation framework has the following characteristics: binding of different classes of transport protocols to the application, dynamic renegotiation of application QOS using the signalling network and, support for multiple resource reservation algorithms.

The paper is organized as follow. In Section 2, we briefly describe the RBG decomposition of the XRM for modeling multimedia networks. In Section 3, we describe the structure of the G-model, and in Section 4, details of its realization as a broadband kernel are presented. In Section 5 we illustrate how to create a multimedia service using broadband kernel services. Finally we conclude with Section 6.

2. RGB Decomposition of the XRM

The XRM models the communications architecture of networking and multimedia computing platforms. It consists of 3 components called, the *Broadband Network*, the *Multimedia Network* and the *Services and Applications Network* (see Figure 1). The broadband network is defined as the physical network that consists of switching and communication equipment and multimedia end-devices. Upon this physical infrastructure, resides the multimedia network whose primary function is to provide the middleware support needed to realize services with end-to-end QOS guarantees over the physical media-unaware network. This is achieved by building upon a set of QOS abstractions derived from the broadband network. This set of QOS abstractions jointly define the *resource management and control space*. The process of service creation calls for resource reservation and distributed state manipulation algorithms. From this perspective, the multimedia network provides a *programming* model that allows service behavior to be specified and executed. Service abstractions represent the states of a service created using algorithms native to the multimedia network. These abstractions are used by the services and applications network for managing and creating new services through dynamic composition and binding.

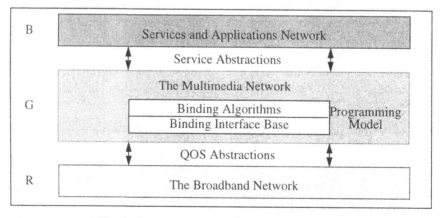

Fig. 1. Overview of the RGB Decomposition of the XRM

The RGB decomposition represents detailed viewpoints of the broadband network, the multimedia network and the services and applications network, respectively. The interface between R-, and G-models is a set of QOS abstractions typically structured as graphs that quantitatively represent various resources in the physical network. The G-model uses these graphs for creating service abstractions that are provided to the B-model for building more complex services. Thus, the interface between the R- and G-models and the interface between the G- and B-models provide abstractions that are similar in structure but differ in usage. For more details on the RGB decomposition please refer to [2], [3] and references therein.

3. The Structure of the G-Model

The role of the G-model is one of creating services. Our notion of G-model services encompasses network services such as virtual networks as well as the low level support functions needed to realize such a service. High level network services are realized within the resource management and control space offered by the R-model. Figure 2 illustrates some examples of services provided by the G-model. These services are realized as objects (algorithms and data structures) and represent what we call the *broadband kernel services* of multimedia networks. They can be used as a building block by an application in the B-model to create multimedia services.

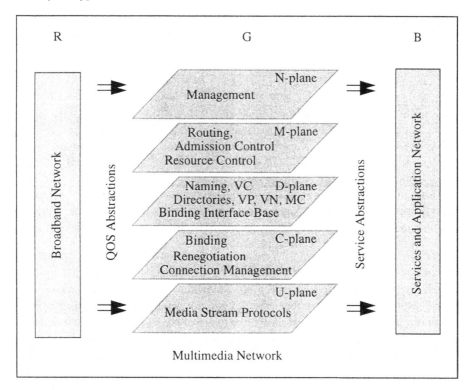

Fig. 2. The G-Model of the XRM

As illustrated in Figure 2, the G-model is divided into five planes. The states of the broadband kernel are stored in the D-plane. The algorithms acting upon them reside in the N-, M-, C- and U-planes. The overall model is known as the XRM; the reader is referred to [1], [2] for details. Although resource control (M-plane) services (routing, admission control, *etc.*) and management (N-plane) are important for the deployment of broadband networks, we do not address the realization of the N- and M-planes here. In this paper, we focus on the realization of the C-plane (connection management) and the U-plane (information transport), and we describe the service creation process.

3.1 Connection Management and Renegotiation (C-plane)

Connection management and renegotiation algorithms resides on the C-plane. A Connection Manager is the coordinator that enables the creation of connection services. The following functionalities are required by any connection manager to perform its task: *route selection, resource reservation, states saving,* and *renegotiation.* We described how these functionalities should be designed so that a large class of connection management schemes can be implemented. A specific call connection model is not prescribed. Instead, multiple connection schemes are envisioned to run simultaneously in the network.

The route selection approach used depends on the routing strategy. Routing is a control algorithm running on the M-plane. The path of a connection is provided by *route* objects in the BIB, whose routes are updated by the *router* objects. There are two extreme route selection approaches: source routing and hop-by-hop routing. In source routing, the *route* object completely specifies a route at once. In hop-by-hop routing, the route object provides sufficient information to progress to the next hop only. In between these two approaches, any combination is possible. Domain routing falls under this hybrid category, where a number of routers are needed to completely specify a route. For example, the PNNI [16] approach to routing can be explained in terms of a high-level router that specifies a set of intermediate hops and the sub-domain routers cooperate to specify the rest. The difference between these approaches is the 'length' of the segment of the entire path provided in an access to the *route* object. The *route* object has to provide the suggested segment of the path and a handle to the next *route* object if the last node on the suggested path is not the destination node. Thus, by accessing the appropriate route objects (in the BIB), the entire path of the connection can be constructed using different approaches.

Resource reservation performed by a connection manager can be divided into two groups: reserving system resources (buffer, bandwidth, CPU cycles, *etc.*), and reserving and setting of identifiers in the network fabric for cell transport. The reservation of system resources should be based on abstractions that are independent of the details of the system hardware and that provide QOS guarantees. For manipulation of switching identifiers, the primitive should be as close as possible to the hardware abstractions so that maximum flexibility is maintained. In earlier work, [3], we have addressed both issues. Our approach to reserving and setting resources is to make accessible the state of the hardware and to allow its modification through a set of generic ATM switch interfaces. In addition, the requests received by the connection manager are specified in terms of application level terminology. In order to reserve (or change) the resource usage, the connection manager needs to be able to specify these requirements in the appropriate terminology. For example, QOS abstractions for the network resources can be defined in terms of calls of predefined class of service; where each of these classes are defined with specific cell loss, cell delay, *etc.* On the other hand, the multimedia network service abstractions are specified in terms of frame rate, frame loss, *etc.* In our framework, a QOS mapper is required for translating QOS specifications between the various abstractions.

Once a connection has been setup, its state must be kept. We believe that the state of a connection (*e.g.*, the bandwidth reserved and the route) must be kept in the switches to make the system more robust. For example, if the connection manager crashes, the manager can be restarted and requested to find out all the routes it established. In our approach, each switch keeps sufficient information so that a connection manager can discover the entire route of a connection by tracing forward and/or backward starting from any hop in the connection. However, such an approach causes heavy interactions among objects during setup, renegotiation and deletion. In order to improve the signalling performance, we allow connection managers to cache the connections state information. In this case, connection managers must assume that the cached information can become invalid. This might arise, for example, if a connection is released by another connection manager.

Renegotiating means asking for more or less system resources for a connection than the one currently committed. Renegotiation may be performed either when the application using the connection explicitly requests it or when it is triggered by the QOS monitoring system due to sustained QOS violations. In [11], renegotiation is also performed during the connection setup phase. In order to efficiently perform renegotiations, resource interfaces must allow immediate changes in resource reservation, *i.e.*, without the need to release the resource first. Furthermore, renegotiation has the potential to overload the signalling system much faster than anticipated from the call request rate. We are currently looking into architectural concepts that can help improving the scalability and performance of the signalling system. Finally, a change in the size of a resource capacity can affect many connections simultaneously; we are also looking into distributed algorithms that can coordinate the resulting renegotiations without causing a signalling overload.

3.2 Information Transport (U-plane)

The information transport, or U-plane, focuses on *QOS-aware transport protocols.* These contain mechanism for processing of media streams (*i.e.*, in-flow and on a fast time scale). For guaranteeing end-to-end QOS, *QOS-based transport APIs* (for provisioning, QOS control and media transfer) are needed and are briefly addressed. Finally, slow time scale *QOS monitoring* is also addressed. Before discussing these service functionalities, we describe the multimedia network media stream services offered at the G- and B-models interface (G‖B).

The U-plane offers media stream services such as motion JPEG, MPEG video and CD quality audio for real-time multimedia applications. For non-real time applications, services such as reliable (error free) and best effort are offered. At the G‖B interface, the QOS is specified in the application level terminology and per session. The QOS parameters are expressed in terms of frames or packets; that is, for real-time services, the frame loss rate, frame gap loss, frame delay, frame peak rate and the maximum frame size, and for non real-time services, the average throughput, average packet delay, maximum packet rate and the maximum packet size. Only losses and delays represent QOS parameters. These are monitored by the receivers. The other parameters are traffic descriptors used for regulating the media flows at the senders.

To support applications with QOS requirements, QOS-aware transport protocols are needed. In particular, the transport protocols need to provide mechanisms to support flow (and rate) control for real-time and non-real time media streams as well as mechanisms to handle error control. The transport protocols should also have the capability to perform in-flow QOS monitoring; that is, measuring frame delay, frame rate, frame loss, *etc.* When fast time scale QOS violations are detected, the protocol should have the capability to adapt; for example, by changing the size of the transport protocol data unit or by reducing its peak rate. Furthermore, the transport protocol should have the capability to the notify network management system or the application if the network does not provided the guaranteed QOS or if a connection is lost.

A QOS-based transport API is needed for provisioning, control and media transfer. Provisioning needs to consider multicast as well as unicast connections. Control APIs are for renegotiation and monitoring. These should permit the network monitoring service to retrieve the in-flow QOS measurements and allow kernel services to set new parameters for flow control, monitoring or violation detection. Finally, the media transfer APIs should provide the transport protocol with the type of information it carries. The reader is referred to [9] for more details.

QOS monitoring is needed for ensuring that the QOS associated with a service is as guaranteed at admission time, for collecting data for management, and for detecting QOS violations and initiating renegotiations. Monitoring needs to be performed on a fast time scale (or in-flow) for flow control mechanisms to adapt to rapid network fluctuations, and on a slow time scale for renegotiations and management purposes. In-flow monitoring is performed in the transport protocol since it is tightly coupled with the media stream. It is a U-plane functionality. Slow time scale monitoring (an M-plane service) can be performed by polling the transport protocol of each active connection for their current measurements or by waiting for transport protocols to forward their measurements. The monitored data needs to be stored and/or processed, and be accessible for management. If QOS violations are detected (*e.g.*, if sustained violations occur), the monitoring service should have the ability to trigger the renegotiation.

To satisfy the above requirements, several components were designed and implemented: **qStack**, a transport protocol for real-time communications that performs in-flow QOS monitoring on a fast time scale; TP, a transport class with a QOS-based API that supports multiple transport protocol suites; and finally TC, a transport controller responsible for the slow time scale operations such as long-term QOS monitoring, QOS violation detection and QOS renegotiation.

3.3 Creating Broadband Kernel Services

In the previous sections, we described the functionality and algorithms that populate the U- and C- planes of the G-model. Because of the rudimentary nature of these services, we likened them to the system services provided by operating systems so that higher level services or applications can be easily realized. In this perspective, we call these algorithms *Broadband Kernel Services* implying that these form a small set of core services that lie at the heart of an extended machine (a broadband operating system) that spans the entire network as opposed to conventional operating systems that reside on a single host. In this section, we give some insight into the structure of broadband kernel services and describe a methodology which allows multiple such services to be composed for creating higher level services. We call this the process of service creation.

The distributed nature of multimedia networks, and therefore of the BIB, suggests many possible means of distributed interactions among different binding algorithms during the service creation process. For example, in a teleconference service, binding algorithms used for connection set up, distributed systems implementing synchronization protocols, resource allocation protocols such as routing, *etc.*, interact cooperatively for realizing the service. The structure of this interaction by and large depends on the higher level service to be constructed. However, sufficient similarities exist between high level multimedia services so that a generalized model of service creation can be inferred. This is described below.

3.3.1 Multimedia Service Creation using Broadband Kernel Services

The generation process is executed by a *Service Provisioning Entity*. Multiple such entities execute in parallel and in a distributed fashion. The service creation process consists of five steps:

1. The creation of a service skeleton for an application (*e.g.*, virtual circuit, virtual path, virtual network or multicast). For example, the skeleton for a virtual circuit consists of a graph from a source node to a destination node.
2. The mapping of the skeleton into the appropriate name and resource space, thereby creating a *network application*.
3. The association (or binding) to the application of a media transport protocol, thereby creating a *transport application*.
4. The binding of the transport application to resources, creating a *network service*.
5. Finally, the binding of the service management system to the network service, thereby creating a *managed service*.

The skeleton is created using name and resource mapping services. The resulting network application is bound to a transport protocol, resources and service management.

When a service is required, the application process issues a service request to the responsible service provisioning entity that in response will invoke the corresponding binding algorithm(s) for establishing a service instance.

This simple view of the service creation process is surprisingly complete. For example, familiar algorithms such as routing belong to step 2, connection management and admission control to step 3, and transport protocol selection to step 4 of the above process. In the next section, we describe a general framework for building servers that provide the services we have just described.

3.3.2 The Structure of Servers and Services

Broadband kernel services and indeed all services are offered by servers which support the creation and maintenance of their state. When a request for a service is made to a server, a service instance is instantiated or created. Each service instance is composed of an algorithmic part and a data part. The algorithmic part expresses the logic of the service instance while the data portion is an abstraction of its state. Individual service instances can be customized by modifying the logic of its algorithmic part. Similarly control can be effected on an executing service instance by the modification of its states. Servers also have an algorithmic component which specifies how service instances are created, deployed and managed and a data part which models the state of the server and its policies.

The interfaces of a server reflect upon the roles that it must play in the process of provisioning a service. Typically these include creation, control, management and programming. Thus, we define 4 general interfaces called the *Service Factory* interface, the *Service Programming* interface, the *Service Control* interface and the *Service Management* interface respectively. The service factory interface is used to request the creation of a service instance. The service programming interface allows customization or modifications to be made to the algorithmic component of the server or service instance. The service control interface is the operational interface to the service instance and allows the monitoring and manipulation of service instance states during execution. Finally the service management interface allows for monitoring and control of the server and the setting of management policies. .

Figure 3 illustrates a server of a high level multimedia service that has 4 interfaces. Recall from discussions in the previous section that the service creation process of such a service requires 5 steps. These 5 steps (shown as a bubble in the figure) are captured as instructions in the script or algorithmic component of the server and are executed during the creation of a service instance. While it is not the case in the figure, it is also possible for some services to have no state (*e.g.*, a simple database lookup service). Such services do not create instances and have only control, management and logic interfaces.

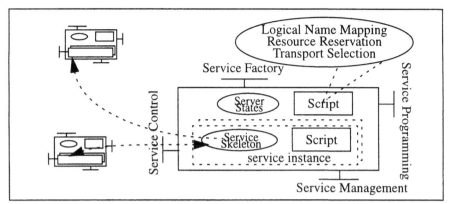

Fig. 3. Interfaces for Multimedia Distribution

4. Prototype Implementation of the Broadband Kernel: xbind 2.0

In this section, we describe **xbind** 2.0, a prototype implementation of the broadband kernel. **xbind** 2.0 is our second generation middleware toolkit. It allows complex multimedia services to be readily assembled by binding together a number of simple objects. It is implemented in CORBA and uses Iona Technologies Orbix 2.0 [8] as the distributed computing platform. We begin by describing the high level system architecture following descriptions of the major system services. Built on top of this toolkit, in Section 5, we present the implementation of a multi-party teleconferencing service that supports dynamic bandwidth renegotiation, per-connection selectable transport and real-time QOS monitoring and control functionality.

Figure 4 gives the high level system architecture of **xbind** 2.0. Four functional planes and the important objects on each plane are highlighted. The information transport, or G::U-plane, contains objects that handle transport activities, *e.g.,* transport protocols and stacks. Objects on this plane interact with primitives of the ATM protocol stacks (*e.g.,* AAL5) in the R::U-plane. Above the G::U-plane, the interactions among objects are on the signalling level. The Binding Interface Base (G::D) is a collection of CORBA interfaces that offers an abstract view of resources in the Binding Architecture. These include naming resources like ATM Virtual Circuit Identifiers (VCIs) or physical resources like multimedia devices. Calls are made to these interfaces to bind the underlying resources to create low level services, *i.e.,* broadband kernel services. Upon the BIB, network services are built on the G::C and G::M-planes. Examples of services in these planes include routing, resource reservation, device management and transport control. These services provide service abstractions and support to the B-model for building more complex multimedia services. In this section, we provide implementation description of objects in the U-, C-, D- and M-planes of the G-model.

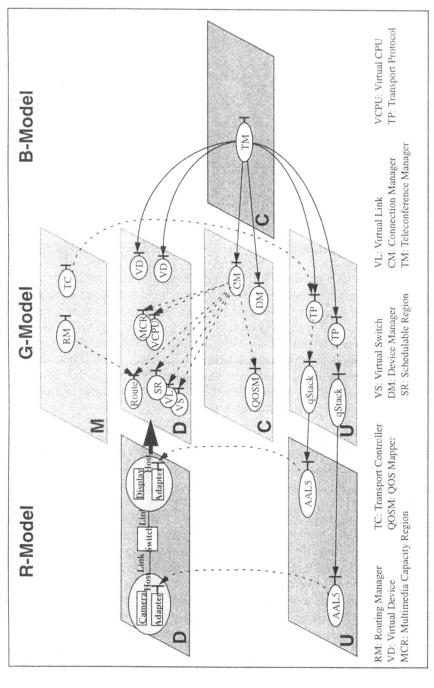

Fig. 4. System Architecture

RM: Routing Manager   TC: Transport Controller   VS: Virtual Switch   VL: Virtual Link   VCPU: Virtual CPU
VD: Virtual Device   QOSM: QOS Mapper   DM: Device Manager   CM: Connection Manager   TP: Transport Protocol
MCR: Multimedia Capacity Region   SR: Schedulable Region   TM: Teleconference Manager

4.1 Media Stream Protocols (U-plane)

For the U-plane, we implemented a transport protocol object with a QOS-based API that supports multiple transport protocol suites and **qStack**, a transport protocol for real-time communications that performs in-flow QOS monitoring.

Transport Protocol (TP)

TP is implemented as a base class that can be used for the development of devices (*e.g.*, camera, speaker, *etc.*). The purpose of TP is to shield the application developer from the complexity of QOS control and management. TP is designed with a well-defined CORBA-IDL interface based on a QOS-based transport API proposed in [9]. TP's API is simple; it has calls for provisioning, control (renegotiation, monitoring and violation notification) and media transfer. It supports both unicast and multicast connections. The transport protocol suites supported by TP are implemented as libraries linked with the applications. The protocols currently supported are **qStack** (see below) and **kStack** [17], the user space implementation of a native-mode ATM protocol stack [14]. We are currently looking into the possibility of supporting RTP/RTCP [15]. In short, TP is a containment object that permits us to support many transport protocol suites transparently.

For each device, TP monitors the QOS received by all of its established connections. It does so by polling the transport protocol stacks for QOS measurements at regular intervals. For each connection monitored, it maintains short and long-term statistics for the detection of QOS violations. Based on its statistics, and some simple rules, it performs QOS violation detection. When QOS violations occur, TP notifies the transport controller (TC) via signalling. When sustained QOS violations occur, the application can be notified via an upstream call.

qStack*: A QOS-aware Transport Protocol*

qStack is a light, real-time transport protocol that performs in-flow QOS monitoring at the frame level (end-to-end). A handshaking mechanism for synchronizing the clocks, negotiating the transport protocol data unit size and the initial sequence number is provided. The clock synchronization mechanism is required for the end-to-end frame delay measurements. **qStack** was originally developed for real-time applications using the AAL5/ATM protocol suite. It can also be used with UDP/IP. The adaptation mechanisms currently supported are limited; the protocol can only adapt the frame rate and peak rate at which the sender transmits the frames. We are currently looking into more sophisticated adaptation mechanisms.

For in-flow QOS monitoring, a sequence number and a time stamp are added to each frame. For each QOS parameter, the minimum, maximum and average values are monitored. On the fast time scale, the averages are estimated over a measurement era of N frames. For slow time scale monitoring these measurements are polled at regular intervals by TP and accumulated for statistics. Overlapping windows are used to perform accurate QOS violation detection. At the end of each measurement window, if the QOS measurements show that the QOS is violated, a notification is sent to the sender via the feedback channel, or to TC via signalling depending upon the time scale on which the of QOS violation occurs. For example, feedback is used for adjusting the frame rate whereas the signalling system is used for notifying the network that too many losses occurred.

To summarize for the information transport, we implmented TP, a transport protocol object with a well-defined QOS-based API that maps the various transport calls to the appropriate transport protocol suite (or stack). We also implemented transport protocol stacks performing in-flow (fast time scale) QOS monitoring at the application level (end-to-end) and flow control (rate control) with QOS constraints.

4.2 Binding Interface Base (D-plane)

The BIB contains many QOS abstractions upon which C-, U- and M-planes act. In this section we described the components implemented as part of **xbind 2.0** only.

Virtual Switch (VS)

The VirtualSwitch interface abstracts the states of VPI/VCI translation tables of an ATM switch. Setting up a path through the switch fabric is achieved by manipulating the physical entries represented by this interface. We also use the VirtualSwitch interface to abstract adapter cards at the end hosts. In this case, the only role played by the interface is in the allocation or reservation of originating and terminating VPI/VCI entries.

Virtual Link (VL)

The VirtualLink interface abstracts the notion of a port in an ATM switch or end host. The primary function of this interface is to associate a notion of switching (or transmission) capacity to a port on a switch (or end host). This interface is used together with the SchedulableRegion interface (see below) to represent the concept of networking capacity at multiplexing points.

Virtual CPU (VCPU)

The VirtualCPU interface abstracts the CPU resource of an end host. It is the analogue of the VirtualLink interface on a switch. It is used to associate processing (as opposed to transmission) capacity to a media processor on an end host. Just like the Virtual-Link, the VirtualCPU interface is used together with the MultimediaCapacityRegion interface (see below) to represent the concept of processing power at multiplexing points in the end host.

The next two interfaces are derived from a base interface called the VirtualCapacityRegion interface used to generically represent the capacity of resources.

Schedulable Region (SR)

The SchedulableRegion interface abstracts the notion of call level capacity. It is usually associated with a VirtualLink and gives a representation of the networking capacity of a port in terms of the number of calls of various classes that can be supported without violating the QOS constraints of each class.

Multimedia Capacity Region (MCR)

The MultimediaCapacityRegion of a VirtualCPU is the analogue of the SchedulableRegion of a VirtualLink. It abstracts the notion of the call level capacity of a media processor in terms of the number of calls of various multimedia network service classes that the processor can support without violating the QOS constraints of each class.

Virtual Device (VD)

The end system devices are abstracted as virtual devices. The VirtualDevice interface abstracts the functional interface of a multimedia device as either a source or a sink of multimedia streams. The interface allows adding, removing, pausing and resuming of streams as well as command parsing for alternating the stream generation process (*e.g.,* changing the compression parameters of a video codec).

Example of currently implemented VDs are the camera, the microphone, the speaker and display devices. We are currently looking into the implementation of recording and playing devices for audio and video as well as a whiteboard device.

Route Object (Route)

The route object contains routes between any two end-points in its domain. The routes it contains are computed by the route manager and downloaded regularly. A route is obtained by specifying two endpoints and the end-to-end delay constraint.

4.3 Connection Management (C-plane)

Connection Manager (CM)

The connection manager is responsible for connection-related operations (adding, removing a connection, renegotiation of resources, *etc.*). To date only unicast connections have been considered for implementation. In order for the connection manager to perform its task, it needs the services of a route manger which provide a list of possible routes between the source and destination node; a QOS mapper that translates resource requirements between transport level and network level and vice versa (fragments of the IDL interfaces defined are given in the Appendix); and a node server which exports all the necessary interfaces for setting up switch routing tables and reserving resources. The scope of the connection manager is end-to-end. It communicates with all the switches along the route, but also with the host machines on both ends of a connection to verify if the end system has the resources to support the connection. As multiple protocol stacks are supported on the host machines, the end-point data structures used is a union that encapsulates data structures for different type of end-points (see Appendix). Upon successful completion of a connection setup, the connection manager returns a network endpoint data structure that contains the VPI/VCIs to be used by the application (or TM).

A connection manager can reside on any host computer. We have implemented both "stateful" and "stateless" connection managers. We allow the addition, removal or renegotiation of a connection to be performed by different connection managers.

QOS Mapper (QOSM)

The QOS mapper is responsible for mapping the application level QOS into network level QOS in terms of COMET classes. All video services are mapped into class I, audio into class II and reliable data into class III. For details on the class mapping, please see [9]. For parameter mapping, the current mapper only does simple rescaling of parameters based on the ratio of the maximum frame to cell size. Furthermore, the mapper currently supports mapping only between network and transport. Ongoing work is currently underway for realizing application-to-transport and user-to-application QOS mapping.

Device Manager (DM)

The device manager is responsible for tracking the capabilities of all multimedia devices on a particular host. Currently four categories of multimedia devices are supported. These are audio input devices such as microphones, audio output devices such as speakers, video input devices such as cameras and video output devices such as displays. Each of these devices in turn may support a number of media formats, rates or options. For example, a camera device may support a particular kind of video encoding format at a particular window size from a particular video source. When a device initializes during the bootup process, it registers with its (device) manager and passes to it a sequence of data structures describing all the formats, rates and options it supports. Thus at any time, the device manager is aware of the details of each device available on its host. Before a device terminates, e.g., due to an error, it attempts to deregister itself from the device manager. The device manager on its own accord polls the devices on a regular basis to ascertain if they are still available.

4.4 Resource Control (M-plane)

Route Manager (RM)

The route manager implements minimum weight routing with delay constraints. The route manager keeps a data structure that contains the network topology and for each link, its weight and delay. Various weight parameters have been experimented with on a prototype implementation (hop count, bandwidth, and utilization, *etc.*). If there is no route between two hosts where the sum of link delays is less than the delay constraint, the two hosts are considered unreachable from each other, even though link bandwidth may be available. The route manager implemented is passive. It only recomputes a new set of routes on request. There can be multiple copies of the route manager running simultaneously. Once the routes are computed, they are stored in the route object.

Transport Controller (TC)

The transport controller implements QOS monitoring on the slow time scale. There is one TC per end system. Every five seconds, it polls every connection it is responsible of monitoring (using TP::query) and gathers network management statistics. Upon sustained QOS violation, it notifies the appropriate service manager (the conference manager in the application example of Section 5). The service manager can then decide to initiate QOS renegotiation or to ignore the notification based on some control rule it may have. TC also monitors each connection to see whether it is active or not. If a connection has not been responding to its queries for more than two minutes, it assumes that the application (or device) crashed.

5. Creating a Multimedia Service from Broadband Kernel Services

To validate our architecture, we implemented a teleconference manager that is built upon the broadband kernel services described in the previous section. Figure 5 shows how the teleconference manager (TM) interacts with the objects in the G::M-, G::C-, G::D- and G::U-planes.

The TM resides on the B::C-plane. It is located in the B-model since it is a multimedia service and resides on the C-plane since its primary function is to create connections to transport multimedia streams. Upon receiving a request from a user through its factory interface, the TM creates an instance of the service skeleton. It then queries each hosts DM involved in the conference to see if any devices capable of supporting the requested type of media stream exist. If successful, a connection setup request is made to the CM who in turns queries the QOSM for a translation between the transport level QOS specified by the BIB components and the network level QOS understood by the VS and VL interfaces. The CM gets its route from the Route object that contains the latest set of routes computed by the RM. Once a route has been obtained, the CM proceeds to each host (VSs/VLs) along the route and reserves the required resources. The connection setup process completes with the CM returning back to the TM a pair of VPI/VCIs representing the entry and address points to the connection that it has established. At this point, the TM proceeds to inform the two TPs at both endpoints to open the associated network interface device using the VPI/VCI pair and transport protocol of choice and gets in return a unique connection identifier that it then passes to the devices (VDs). The connection identifier serves to abstract away any details about the transport protocol (such as the actual VPI/VCI pair or transport stack in use) from devices so that different simultaneous transport protocol stacks can be used by the same device without any change required. Once this is completed, a session is said to have been established and a multimedia stream generated at the source device can be carried over the network to the destination device.

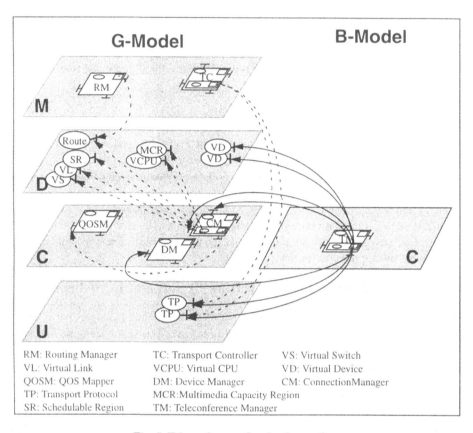

Fig. 5. Teleconference Service Invocations

During the lifetime of the connection, the TC monitors the QOS obtained in the end system for any violation of QOS. In the situation where sustained QOS violations occur, the TC may initiate a renegotiation request to the TM who would in turn request the TP of the transmitting device to reduce its rate. The TP could in turn interact with the device to effect the rate control. Alternatively, if a feedback channel to the transmitting device is available, as maybe required by some protocols, then the TP of the receiving device can directly notify the TP of the transmitting device of the violations. Still another alternative is when the user requests for a change in the grade of service. In this case, the TM needs to issue a renegotiation request to the CM who then attempts to renegotiate for new resources via the RM and QOSM on the respective hosts. The other interfaces (management and programmability) of the TM exist so that other B::N-plane applications can monitor and control service creation policies of the TM on-the-fly.

The Client Application: A Java-CORBA GUI Teleconferencing Service Applet

A user interface for the teleconferencing service was implemented. Figure 6 shows a screen dump of it. Essentially, the interface is realized as a single Java applet with communications to the back-end CORBA-based teleconference manager achieved using Iona Technologies' OrbixWeb [7] toolkit. The toolkit includes a compiler that automatically generates a Java proxy class from an IDL interface and a set of Java libraries that implement the Internet Inter-ORB Protocol (IIOP) feature of CORBA 2.0 for applet-to-ORB communication. The role of the applet is thus merely to collect user input and to make the appropriate CORBA invocations to the teleconference manager when the user confirms a selection. As there is currently no simple way of allowing communication in the reverse direction (*i.e.*, for the back-end server to call the front-end applet), the relationship between the applet and CORBA server is currently restricted to only client and server, respectively.

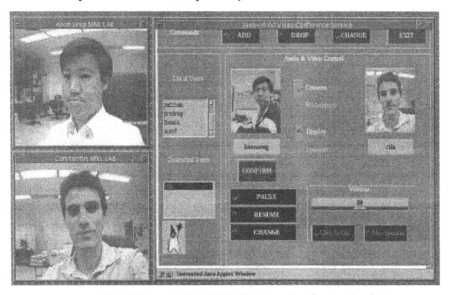

Fig. 6. Figure 2. xbind Tele-Conferencing Service Applet

6. Conclusions and Future Work

There are two key defining functional characteristics of our broadband kernel: an extended machine and a resource manager. By focussing our current implementation on the extended machine characteristics, we have created the first programmable platform for realizing multimedia services on ATM-based broadband networks.

We have also described a methodology of service creation and showed how broadband kernel services can be used for the establishment of network services. Experience with the broadband kernel prototype, including measurements and performance, will be reported in the future under URL: http://www.ctr.columbia.edu/comet/xbind.

7. References

[1]   Lazar, A.A., Bhonsle, S. and Lim, K.S., "A Binding Architecture for Multimedia Networks", *Journal of Parallel and Distributed Computing*, Vol. 30, No. 2, Nov. 1995, pp.
204-216.

[2]   Lazar, A.A., Lim, K.S. and Marconcini, F., "Realizing a Foundation for Programmability
of ATM Networks with the Binding Architecture", *IEEE Journal of Selected Areas in
Communications*, Special Issue on Distributed Multimedia Systems, Vol. 14, No. 7, September 1996.

[3]   Project **xbind**, http://www.ctr.columbia.edu/comet/xbind/xbind.html.

[4]   OMG, The Common Object Request Broker: Architecture and Specification, Rev. 1.2,
Dec. 1993.

[5]   TINA-C, Service Architecture Version 2.0, Document No. TB_MDC.012_2.0_94, Mar.
1995.

[6]   Barr, W. J., Boyd, T. and Inoue, Y., "The TINA Initiative," *IEEE Communications Mag.*,
Mar. 1993.

[7]   Iona Technologies Ltd., Orbix for Java, *White Paper*, Cambridge MA., Feb. 1996.

[8]   Iona Technologies Ltd., Programmers Guide, Orbix 2.0, *http://www.iona.ie*, July 1995.

[9]   Huard, J.-F., Inoue, I., Lazar, A.A., and Yamanaka, H., "Meeting QOS Guarantees by
End-to-End QOS Monitoring and Adaptation." In *Proceedings of the Fifth International
Symposium on High Performance Distributed Computing (HPDC-5)*, Syracuse, NY,
Aug. 1996.

[10]   Banerjea, A. Ferrari, D. Mah, B.A., Moran, M., Verma, D.C., and Zhang, H., "The Tenet
Real-time Protocol SuiteL Design Implementation and Experiences." *IEEE/ACM Trans.
Networking*, Vol. 4, No. 1, Feb., 1996.

[11]   Parris, C., Zhang, H., and Ferrari, D., "Dynamic Management of Guaranteed-Perfor-
mance Multimedia Connections." *Multimedia Systems*, No. 1, 1994, pp. 267-283.

[12]   Chan, M.C., Pacifici G., and Stadler, R., "Prototyping network architectures on a super-
computer." In *Proceedings of the Fifth International Symposium on High Performance
Distributed Computing (HPDC-5)*, Syracuse, NY, Aug.1996.

[13]   Chan, M.C., Hadama, H. and Stadler. R., *"An Architecture for Broadband Virtual Networks under Customer Control", IEEE Network Operations and Management Symposium (NOMS'96), April 1996, Kyoto, Japan.*

[14]   Ahuja, R., Keshav, and S., Saran, H. "Design, Implementation and Performance of a
Native Mode ATM Transport Layer." In *Proc. of IEEE Infocom*, San Francisco, CA, Mar.
1996. Extended version to appear in *IEEE/ACM Trans. on Networking*.

[15]   Schulzrinne, H. G., Casner, S., Frederick, R. and Jacobson, V., "RTP: A Transport Protocol for Real-Time Applications," *IETF RFC 1889*, Jan. 1996.

[16]   ATM Forum, "PNNI: Private Network-Node Interface Specification Version 1.0." March
1996.

[17]   **kStack**: http://www.ctr.columbia.edu/comet/software/kStack.

8. Appendix

8.1 IDL Fragments from "QOS.idl"

```
BIBStatus QOSmap(
     in QOSSpecification from,
     inout QOSSpecification to) raises (Reject);

union QOSSpecification switch(Level)
{
     case user:         UserQOSSpecification        u;
     case application: ApplicationQOSSpecification a;
     case transport:   TransportQOSSpecification   t;
     case network:     NetworkQOSSpecification     n;

};
struct TransportQOSSpecification
{
     TransportService  service;
     TransportQOS      qos;
     TransportProtocol protocol;
};
struct TransportQOS {                 // per frame
     float delay;
     float loss;
     short gap_loss;
     float PDU_peak_rate;
     float PDU_max_size;
};
```

8.2 IDL Fragments from "EndPoint.idl"

```
struct EndPoint {
     EPFormat epformat;
     char hostname[128];
     EndPointId epid;                 // data channel
     EndPointId fepid;                // feedback channel
     Direction  dir;
};
union EndPointId switch(EPFormat)
{
     case IP_epf:       IPEndPointId    ip;
     case ATM_epf:      ATMEndPointId   atm;
     case E164_epf:     E164EndPointId  e164;
     case CORBA_epf:    CORBAEndPointId corba;
};
struct ATMEndPointId {
     short port;
     short VPI;
     long VCI;
};
```

Specifying QoS for Multimedia Communications within Distributed Programming Environments

Daniel G. Waddington, Geoff Coulson and David Hutchison

Computing Department,
Lancaster University,
Lancaster LA1 4YR
e-mail: [dan,geoff,dh] @comp.lancs.ac.uk

Abstract

Because of the increasing emphasis on distributed object programming for the provision of multimedia telecommunications services, it has become apparent that a unification between the distributed programming environment and techniques for QoS specification in multimedia communications must be made. Various QoS frameworks and QoS reference models have already been established and/or standardised. Each has approached the problem of QoS specification in its own way but it is possible to draw similarities across the board. This paper examines four different approaches to QoS specification and attempts to integrate these ideas into the distributed programming environment (DPE).

1. Introduction

QoS can be defined as how "good" a communications service is between two or more points. QoS is hierarchical in that an end-to-end definition of QoS can be represented as a series of point-to-point QoS specifications. A QoS contract is interpreted as a two-way service contract whereby the QoS provider undertakes to support a given level of QoS if and only if the traffic generator undertakes to supply its data at the agreed QoS [Hutchison, 94]. This paper is concerned with the specification of QoS for multimedia communication at the application layer. We propose an approach to QoS specification derived from a study of several previous approaches and then discuss the application of this approach to multimedia telecommunication service provision through Distributed Programming Environments (DPEs). DPEs, which are layered on top of the operating system and communications system, provide an object based computational model for the benefit of programmers of distributed multimedia applications and facilitate flexible inter-object communications in a networked environment.

The rest of this paper is structured as follows. Section 2 introduces concepts relevant to distributed programming whilst section 3 defines various QoS concepts and definitions. Section 4 discusses some accepted QoS frameworks and section 5 outlines techniques of service provision through DPEs. Section 6 examines how the ideas of specifying QoS can be applied in DPEs and, in particular how these ideas could be used in the provision of multimedia network services. Finally section 7 draws our conclusions.

2.   Distributed Programming Environments

The role of a distributed programming environment (DPE) is to provide the application programmer with a set of tools and abstractions to facilitate both location transparent and platform independent inter-object communications. DPEs communicate via message passing in the form of method requests/replies and within a distributed multimedia computing application objects also require the ability to exchange QoS dependent continuous data streams. Many DPEs only offer abstractions for remote method invocations, including CORBA[OMG, 91], however an increasing number of research bodies are currently looking at the provision of continuous media stream interactions [Coulson, 96] [Halteren, 95]. A DPE offers its tools and services to the application programmer through both compile-time libraries and programming tools, together with run-time domain resident server processes.

2.1   Services of a DPE

2.1.1   Simple Object Oriented Programming (OOP) model

Distributed programming models are based usually on OOP techniques. OOP provides a simple model for data and method encapsulation, together with clear units of functionality (i.e. the object and its interface) which are ideally suited to distribution.

2.1.2   Location transparency and object location services

Within a DPE, a client application is able to 'transparently' make an invocation on any object method within its domain[1] as if it were in its own address space. In contrast to the standard programming paradigm, the programmers domain is extended out of the application address space into the network environment. To clarify the distinction between simple Remote Procedure Call (RPC) environments and DPEs, the latter offers additional services for remote object invocation, including object location and client authentication. Object location allows a client to 'enquire' on the availability of objects and their methods within a domain. Client

---

[1] A domain may be physical or logical. A physical domain is comparable to physical communication paths between nodes, whilst a logical domain is analogous to the concept of a virtual network connection. Logical domains may even be a hierarchy of other domains.

authentication services provide mechanisms for ensuring security between end-users and service providers.

In addition to encapsulating the communications for inter-object method calls, bindings may also offer abstractions for continuous media communications between objects. Such services are more complex to abstract upon, since such interactions often require strict temporal dependencies and other QoS requirements.

2.1.3   Platform Independence

DPEs also offer to the application programmer platform independence. This means that a programmer can make a remote method invocation, or process a media stream without concern to the remote platform. This is achieved through the use of a universally accepted language to define an objects interface signature (an interface is a window onto the methods or the interactions offered by a particular object instance).

2.2   Common Object Request Broker Architecture (CORBA)

CORBA is the distributed object model, produced as a result of research of a consortium of vendors called the Object Management Group (OMG). The model which is now in its second version (CORBA 2.0) is one of the accepted standard distributed computing models.

At the core of the CORBA architecture is the ORB itself. The ORB is the object interconnection bus, where clients are insulated from the mechanisms used to communicate with server objects. Within CORBA, the OMG have also specified how to define an interface between a component and the object bus, using the Interface Definition Language (IDL). Components specify in IDL the types of services that they provide, including methods they export, their parameters, attributes, error handlers, and inheritance relationships with other components. IDL becomes the contract that binds client to server components. It is the use of IDL that shields the client application programmer from the details of implementation and the supporting platform of remote server objects.

Other than the provision of inter-object communications, the ORB offers various additional services. These services include object location, run-time interrogation of remote object interfaces, client authentication, object naming, event notification, concurrency control and more. Finally, the ORB is realised as a set of libraries and tools available to the application programmer.

3.   QoS Concepts and Definitions

QoS specification is concerned with the required levels of quality interpretable in a particular layer of a system. In order to define how QoS is specified we need first to establish some basic terms which are introduced in the following sub-sections.

3.1   QoS Views

QoS is apparent at all layers in a communications architecture but is 'viewed' differently by each layer. A QoS view consists, in general, of notations for quality characterisation, service commitment and cost, together with a protocol for peer to peer QoS negotiation. Figure 1 illustrates the various QoS views of a typical system. The end user is concerned with perceptual QoS which essentially defines how "good" a service appears to the user, together with non-performance related QoS such as cost. User QoS is also the least *service-specific*[2] representation and is produced as a direct effect of application's performance and the supporting environment. There is a fine line between application QoS and user QoS, but it is probably best differentiated by the fact that application QoS is usually in terms of computational concepts where user QoS is not.

System QoS defines the QoS expected by the application of the underlying system. Incidentally, the system incorporates everything between the application and the hardware devices, including the DPE, the Operating System and any reusable system components[3]. System QoS is more service specific as its QoS characteristics (see next section) are directly relevant to the type of service.

Finally, at the lowest level, QoS is specified with respect to device capabilities, either peripheral or network. These QoS specifications, which we term 'device QoS' and 'physical network QoS', are the least service specific. Such specifications are usually very detailed and often concerned with the performance a particular device can offer. The term physical network QoS is simply the device QoS associated with networking, since from an engineering viewpoint there is no real difference between a peripheral device and a physical network device. Physical network device QoS can be characterised in terms of throughput, delay and error rate. However, system services over the physical network devices, such as those provided by the transport layer, are described with more complex system QoS characteristics.

Examples of network and device QoS include ATM cells per second of a virtual connection or video frames per second of a video capture device.

---

[2] Service specificity relates to how dependent upon the type of service the QoS specification is.

[3] System components include available resources such as codecs, filters and device abstractions.

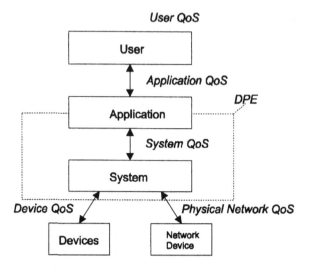

Fig. 1. QoS Views

3.2   QoS Characteristics

A QoS characteristic is a quantifiable aspect of QoS which is defined independently of the means by which it is represented or controlled [ISO, 95]. QoS characteristics are used to formulate a representation of the behaviour of an entity. Because of the diversity of multimedia services, there is no finite set of QoS characteristics. The idea of defining a set of 'generic' QoS characteristics is unsound and can only be realistically applied to particular devices; the concept of resource varies between layers and QoS characteristics are therefore often layer dependent. The process of making hierarchical translations between characteristics of different layers is known as *QoS mapping*. Some QoS characteristics can also be derived, as opposed to mapped, from other characteristics. An example of such derivation is forming the variance of transit delay from transit delay, both are still defined as QoS characteristics. This reinforces the assumption that there is no single finite set of QoS characteristics suited to all requirements.

In addition to deriving characteristics, we can also 'specialise' characteristics. A specialisation may or must be applied in order to make a characteristic concrete and useable in practice [ISO, 95]. Several levels of specialisation can also exist, for example:

> *time delay*
> → *transit delay*
> → *transit delay between Network Service Access Points*

3.3 QoS Categories

Logically related QoS characteristics can be grouped into categories corresponding to a classification of topics relating to QoS at its highest level. Table 1 presents an extensive selection of QoS categories which we have derived from various well defined QoS frameworks (in later sections we look at QoS categories and characteristics in more specific detail). The timeliness and volume categories are of principal concern to the provision of multimedia services, nevertheless other categories including accountability are still important.

QoS Categories	QoS Characteristics
Timeliness	Latency
	Delay jitter
	Recovery time
	Guarantee
	Synchronisation Intervals
	Availability
	Setup time
Volume	Throughput
	Peak Rate
Accuracy	Addressing accuracy (Cell Insertion Ratio)
	Error rate (Cell Loss Ratio, Bit Error Rate …)
	Integrity
Robustness	Reliability (Mean Time Between Failures)
	Maintainability (Mean Time To Repair)
	Resilience
	Survivability
Accountability	Cost
	Auditability
Manageability	Monitorability
	Controllability
Privacy	Authenticatability
	Confidentiality
	Traffic Flow Security

Table 1. QoS Categories and Characteristics

3.4 QoS Contracts and QoS Guarantees

A QoS contract is an agreed QoS specification between one or more entities. Contracts are of particular importance since a component 'C' can only be expected to 'provide' a defined level of QoS if and only if the requested 'required' QoS is also met by other components on which 'C' depends. If the bilaterally agreed QoS specification is not maintained, then the contract is broken. During the negotiation phase of service set up, various QoS contracts must be agreed between the consumer, retailer, network provider and service provider.

Another aspect concerning QoS contracting is the concept of *level of service* or *service guarantee*. The guarantee defines to what extent a party will respect QoS parameters within a contract. Three classes of service guarantee are defined [ISO, 95]: *Compulsory, Statistical reliable (or Guaranteed service)* and *Best effort*. A guarantee level can be assigned to each parameter in a QoS contract or to the contract as a whole.

3.5   QoS Notation

QoS characteristics must be represented in a suitable notation to be of any use. The following describes some of the more common methods of QoS notation.

- *Single Static QoS Parameters* - representation of QoS characteristics through a single static value.

```
jitter = 1ms
throughput = 100kbps
peak duration = 10ms
```

- *Multiple QoS Parameters* - usually in the form of pairs. Multiple QoS parameters enable bounds upon QoS characteristics to be declared and more complex representations such as synchronisation intervals. For instance it may be a QoS requirement that a particular media stream be capable of supporting synchronisation at 10, 20 or 30 millisecond intervals.

```
delay = <10,100>          {min, max}
synchronisation intervals = <10,20,30>
```

- *QoS Structures* - more complex data structures including unions.

```
struct VideoQoS {
        union Throughput switch (Guarantee)  {
                case Statistical: float Mean;
                case Deterministic: float Peak;
                case BestEffort:
                        struct Interval {
                                float min;
                                float max;        };
        };
        union Jitter switch (Guarantee) {
                case Statistical: float Mean;
                case Deterministic: float Peak;
        };
};
```

- *QoS Specification Languages* - such languages provide a more flexible specification mechanism. Examples include QL [Stefani, 93] and SQTL [Lakas, 96]. These notations express QoS as a set of temporal logic formulae that are linearly interpreted on timed state sequences. The formulae are built upon a set of atomic propositions using logical and temporal operators.

example of QL:

```
where 'e' is the event identifier and 'n' the occurrence
of 'e';

  delay_QoS: T(e,n+1) - T(e,n) = 20 ms
  jitter_QoS: 10 < T(e,n) - T(e,n-1) < 100 ms
```

example of SQTL:

```
always capture_frame with_rate 15
if x then play_frame with_rate 25 else
play_frame with_rate 15
```

4. QoS Frameworks

In this section we present QoS frameworks that provide outlines for the concept of quality of service within a particular environment. For instance the ISO QoS framework, as described in section 3.1, is aimed at the provision of QoS in an ISO network environment. The ATM Forum's guidelines on aspects of QoS are aimed at QoS provision in an ATM networking environment and the Lancaster QoS-Architecture attempts to give guidance for aspects of QoS within an ATM and multimedia computing environment.

4.1 ISO QoS Framework

The basic ISO QoS framework is intended to assist those defining communications services and protocols, and those designing and specifying systems, by providing guidance on QoS applicable to systems, services and resources of various kinds. It describes how QoS can be characterised, how QoS requirements can be specified, and how QoS can be managed.

The ISO QoS framework key concepts include:

- *QoS requirements*, which are realised through QoS management and maintenance entities;
- *QoS characteristics*, as previously defined in section 3.2;
- *QoS categories*, which represent a policy governing a group of QoS requirements specific to a particular environment such as time-critical communications (see also section 3.3).

Table 2 shows the QoS categories and characteristics as defined by the framework. Each of these groups is considered to be relatively independent and it is intended that systems implement a specific set of characteristics, according to some QoS policy.

QoS Category	QoS Characteristics
Time-related	delay, validity, remaining lifetime
Coherence	temporal coherence, spatial consistency
Capacity-related	capacity, throughput, processing capacity, operation loading
Integrity-related	accuracy errors
Safety-related	safety
Cost-related	cost, user cost, data cost, resource cost, event cost
Security-related	protection, access control, data protection, confidentiality, authenticity
Reliability-related	availability, reliability, fault containment, fault tolerance, maintainability
Other	precedence

Table 2. ISO QoS Framework Categories

Table 3 expands on the definitions for some of the ISO QoS framework defined characteristics, further details can be gained from [ISO, 95].

Temporal Coherence	Indicates whether an action has been performed on each value in a list within a given time window.
Spatial Consistency	Indicates whether the value of each variable in a list has been consumed in a given time window.
Capacity	The amount of service that can be provided in a specified period of time.
Processing Capacity	An amount of processing performed in a period of time
Operation Loading	Ratio of capacity used to capacity available
Accuracy	Accuracy is a QoS characteristic of concern to the user, for whom this characteristic refers to

	the integrity of the user information only. (The integrity of headers and similar protocol control information may be the subject of other characteristics)
Protection	The security afforded to a resource or to information
Access control	Protection against unauthorised access to a resource
Availability	The proportion of time that satisfactory service is available
Maintainability	Duration of any continuous period for which satisfactory, or tolerable, service is not available, related to some observation period (quantified as the probability Mean Time To Repair)
Fault Containment	Ability to operate in the presence of one or more errors/faults
Precedence	The relative importance of an object or the urgency assigned to an event

Table 3. ISO QoS Characteristics in detail

In addition to QoS characteristics and categories, the framework outlines various aspects of QoS management. The framework is made up of two types of management entity that attempt to meet the QoS requirements by monitoring, maintaining and controlling end-to-end QoS [Campbell, 96]:

i) *layer specific entities:* The task of the policy control function is to determine the policy which applies at a specific layer of the open system. The policy control function models any priority actions that must be performed to control the operation of the layer. The definition of a particular policy is layer-specific and therefore cannot be generalised. Policy may, however, include aspects of security, time-critical communications and resource control. The role of the QoS control function is to determine, select and configure the appropriate protocol entities to meet layer-specific goals.

ii) *system wide entities:* The system management agent is used in conjunction with OSI system management protocols to enable system resources to be remotely managed. The local resource manager represents end-system control of resources. The system QoS control function combines two system-wide capabilities; to tune performance of protocol entities and to modify the capability of remote systems via OSI systems management. The OSI systems management interface is supported by the systems management manager which provides a standard interface to monitor, control and manage end-systems. The system policy control function interacts with each layer-specific policy control function to provide an overall selection of QoS functions and facilities.

The OSI QoS management framework also outlines general QoS mechanisms and their operational phases, the assignment of QoS function to entities and definitions for conformance, consistency and compliance.

4.2 ATM Forum UNI 3.1 Recommendations

The ATM Forum's UNI 3.1 ATM bearer service QoS guidelines are based on the ITU-T I.350 recommendations [ITU, 93]. The aspects of QoS that are covered are restricted to the identification of parameters that can be directly observed and measured at the point at which the service is accessed by the user. The forum assumes that QoS has a direct relationship to the Network Performance (NP) and that the principal difference is that QoS pertains to user oriented performance concerns of an end-to-end service, while NP is concerned with parameters that are of concern to network planning, provisioning and operations activities. To facilitate QoS provisioning in an ATM network, the forum defines ATM performance parameters and various QoS classes.

4.2.1 ATM Performance Parameters

Table 4 summarises the set of ATM cell transfer performance parameters which correspond to the generic criteria of the assessment (shown in the right hand column) of the QoS.

Cell Error Ratio	$$\dfrac{\text{Errored Cells}}{\text{Successfully Transfered Cells} + \text{Errored Cells}}$$	Accuracy
Severely-Errored Cell Block Ratio	$$\dfrac{\text{Severely Errored Cell Blocks}}{\text{Total Transmitted Cell Blocks}}$$	Accuracy
Cell Loss Ratio	$$\dfrac{\text{Lost Cells}}{\text{Total Transmitted Cells}}$$	Depend-ability
Cell Misinsertion Ratio	$$\dfrac{\text{Misinserted Cells}}{\text{Time Interval}}$$	Accuracy
Cell Transfer Delay	Delay in transfer of an ATM cell between nodes.	Speed

Mean Cell Transfer Delay	Mean CTD is defined as the arithmetic average of a specified number of cell transfer delays for one or more connections	Speed
Cell Delay Variation	Describes 1-point Cell Delay Variation (1-point CDV) and 2-point Cell Delay Variation (2-point CDV). 1-point variation is measured with reference to a single measurement point, where as 2-point variation is measured from the output of a connection portion with reference to the pattern of the corresponding events observed at the input to the connection portion.	Speed

Table 4. I.350 ATM Performance Parameters

4.2.2   QoS Classes

UNI 3.1 also establishes the concept of QoS classes. A QoS class can have a specified performance parameter (Specified QoS) or no specified performance parameters (Unspecified QoS). QoS classes are inherently associated with a connection. The network itself may support several specified QoS classes and at most one unspecified QoS class. The performance provided by the network should meet (or exceed) performance parameter objectives of the QoS class requested by the ATM end-point. Both ATM VPCs and VCCs should indicate the requested QoS by a particular class specification.

Specified QoS Parameters

A Specified QoS class provides a quality of service to an ATM virtual connection (VCC or VPC) in terms of a subset of the ATM performance parameters previously described. For each Specified QoS class, there is one specified objective vale for each performance parameter. The following list shows those QoS classes which have been defined in UNI3.1 (the italicised print represents the Service Class).

Specified QoS Class 1: *Circuit Emulation, Constant Bit Rate Video*
   - should yield performance comparable to current digital private line performance.

Specified QoS Class 2: *Variable bit Rate Audio and Video*
   - is intended for packetized video and audio in teleconferencing and multi-media applications.

Specified QoS Class 3: *Connection-Oriented Data Transfer*
   - is intended for interoperation of connection oriented protocols, such as Frame Relay.

Specified QoS Class 4: *Connectionless Data Transfer*
- is intended for interoperation of connection-less protocols, such as IP, or SMDS.

Unspecified QoS Class

No objective is specified for the performance parameters. However, the network provider may determine a set of internal objectives for the performance parameters. In fact, these internal performance parameter objectives need not be constant during the duration of the call.

4.3 Lancaster's QoS-A

Lancaster's QoS Architecture [Campbell, 94] offers a framework to specify and implement the required performance properties of multimedia applications over high-performance ATM based networks. This architecture does take into account the higher communication layers and incorporates notions of media flows, service contracts and flow management. It also acknowledges the different representations of QoS between layers, including the representation of QoS within a distributed platform. Table 5 represents the QoS characteristics relevant to the QoS-A Multimedia Enhanced Transport Service [Campbell, 94], which has a strong emphasis on full end-to-end guarantees.

QoS Category	QoS Characteristics
Flow specification	media type, frame size, frame rate, burst, peak rate, delay, loss, interval, jitter
Adaptation	loss adaptation, jitter adaptation, throughput adaptation, delay adaptation, disconnect
Maintenance	maintenance
Connection-related	service, start time, end time
Cost	cost

Table 5. QoS-A QoS Categories

Media Type	Common flows for video, voice and data.
Frame Rate/Frame Size	Average throughput requirement.
Peak Rate	Maximal throughput.
Burst	Size of potential bursts of traffic.
Delay	End-to-end delay.
Maintenance	Options for maintenance are monitor, maintain and no_maintain. Monitor instructs the QoS-A to periodically deliver measured performance

	assessments relating to the specified flow. Maintain, on the other hand, attempts to transparently exert fine grained corrective action to maintain QoS levels according to the service contract, but does not deliver any assessments.
Service	Support for fast, negotiated or forward service types. Negotiated service supports full end-to-end protocol negotiation. Fast connect service reservation and data transfer phases coincide. Forward reservation allows network and end-system resources to be booked ahead of time.
Start time/End time	Service window bounds.

Table 6. QoS-A QoS Characteristics in detail

The precise interpretation of the QoS characteristics as defined in Table 5, is determined by the commitment specification. The QoS-A `commitment_t` parameter is the same as the service guarantee as defined in Section 2.4.

Although the QoS-A architecture only specifies a small set of QoS characteristics relevant to media transport the model does realise the necessity for the management of QoS at all levels and the mapping of QoS between the layers in the architecture.

4.4   TINA-C Recommendations

The Telecommunications Information Network Architecture (TINA) Consortium is an international collaboration aiming at defining and validating an open architecture for telecommunications systems for the broadband, multimedia, and information era [TINA, 94]. The intention of the consortium is to make use of recent advances in distributed computing, particularly the advent of the CORBA distributed object model, for the provision of telecommunication services in an open architecture. TINA-C have identified the requirements for QoS specification at both the network and service level. We now briefly discuss their approach to specifying QoS at the service level, since this has a direct relevance to QoS specification in a DPE.

TINA acknowledges the distinction between QoS associated with a contract between objects and their environment (object service attributes) and streams and their environment (stream service attributes). This is further divided into *functional* and *non-functional* QoS requirements; functional requirements which are relevant to the operation/stream signatures and non-functional requirements provided by the DPE, e.g. availability, security. Tables 7 and 8 show some of the TINA object and stream service QoS attributes.

QoS Category	QoS Characteristics
Availability	Peak MTBF, Mean MTBF
Performance	Peak Response Time, Mean Response Time
Others	Security, Consistency, Reliability

Table 7. TINA Object Service Attributes

QoS Category	QoS Characteristics
Throughput	Peak Throughput, Mean Throughput, Bound Throughput
Jitter	Peak Jitter, Mean Jitter
Delay	Peak and Mean Delay
Error Rate	Peak and Mean Error Rates

Table 8. TINA Stream Service Attributes

Because the CORBA object model is essentially limited to simple request/reply method interactions the TINA Consortium extended the model to support multiple interfaces on an object and stream interfaces on an object. Their interface definition language, known as ODL-95, supports these requirements and the necessary placeholders for the specification of QoS. Section 4.2 briefly introduces the use of ODL-95 and further information can be gained from [OMG, 95].

5.   Distributed Multimedia Models

This section is intended to outline the general architecture of distributed multimedia platforms and their requirements for QoS support in the provision of broadband networking services. Distributed multimedia applications are built up through the logical "binding" of multimedia service components [ODP, 92]. For peer-to-peer communications, this binding is usually controlled through a composite *binding object*. The binding object is the client's point of contact for the establishment and control of a multimedia service, see figure 2.

During the service establishment phase, which involves the selection and local binding of various service components, the binding object is responsible for type checking the compatibility of the to-be-bound stream interfaces. This type checking includes the verification of required and provided QoS through the examination of the interface's associated QoS specification.

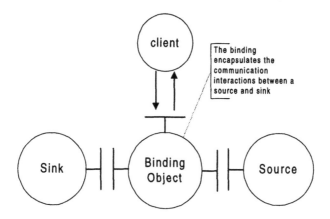

Fig. 2. Binding model

5.1   Operational Interfaces

Operational interfaces comprise a set of object method signatures (a method is defined as a unit of functionality offered by an object). Methods are invoked in a request/reply interaction style and invocations made upon them in a distributed programming environment are both location and implementation transparent[4]. Within a distributed multimedia system operational interfaces are primarily used for control interfaces on system and service objects.

Below is an example standard CORBA operational interface description. The standard CORBA model does not support QoS specification on object methods.

```
interface VideoControl {
        void getBrightness(in short level);
        short setBrightness(void);
        };
```

5.2   Event Interfaces

To specify the continuous flow of media in multimedia applications, it is necessary to create an abstraction which captures the concept of information flowing over time [Blair, 95]. We could model such continuous interaction as asynchronous method calls on operational interfaces, however with this approach there is no concept of a connection governed by overall QoS, as each invocation is a separate isolated event. Therefore to facilitate this requirement we add the concept of *event interfaces*. Event interfaces provide a signature describing unidirectional points of interaction, *events*. An event is described by an identifier, directionality and associated data. Events are also typed.

---

[4] Different object models, such as CORBA, Distributed COM and Distributed SOM each have their own method of achieving this transparency goal.

Event interfaces specify component level continuous interactions, these being interactions which have relationships to time or some other interactions. An event interface as described in Lancaster's extended CORBA computational model [Coulson, 96] describes purely the interaction points without any reference to values of required QoS. The inheritance of QoS_events provides the signature for standard events used for QoS management. These standard events provide mechanisms for feedback from monitoring and notification of QoS degradation or failure.

```
events VideoDecompressor : QoS_events {
        in FrameCompressed(INDEO_T frame);
        out FrameDecompressed(RAW_T frame);
        out DecompressionFailed(MESSAGE_T reason, TIME_T t);
        };
```

Below is described an event interface (known in TINA terminology as a stream interface) using the TINA Object Definition Language [Kitson, 95]. The QoS structures are simple and static.

```
interface VideoCompressor {

    behaviour "This is a Indeo Video Compressor";

    sink VideoFlowType display with Video_QoS requiredQoS;
    sink AudioFlowType speaker with Audio_QoS requiredQoS;
    source VideoFlowType camera with Video_QoS offeredQoS;
    source AudioFlowType mic with Audio_QoS offeredQoS;
    };
```

The most prominent differences between the two above described approaches to event interface specification is that the TINA-ODL description uses the concept of flow type to describe the associated flow of data. Lancaster's Event Definition Language (EDL) allows multiple data types to be associated with a particular event. This approach gives a wider flexibility in the specification of continuous data interactions.

Event interfaces are "locally" [5] bound through the sharing of resources. This binding provides the communication path from source to sink and establishes an overall quality of service associated with the connection. Before the event interfaces are bound, they are typed checked to ensure that each interface provides the QoS as required by the adjacent interface. Such type checking is an important part of QoS management, however the issue will not be discussed further in this paper.

---

[5] Local binding means that there is no stub or skeleton proxy code involvement. The stub provides client side location transparency and the skeleton provides server side location transparency support.

6. QoS Specification on Interactions within a DPE

In this section we extend the ideas of QoS specification described in section 4 and propose an approach to the application of QoS within a DPE. We define the concept of a QoS interaction point, which is any interaction between entities which may require the specification of QoS constraints. Each of the frameworks and recommendations as previously described, are concerned with specifying QoS at one or more of these interaction points. Both the ISO QoS framework and the ATM Forum's recommendations emphasise QoS parameters which are relevant to interactions on the underlying network, whilst in addition to this, the Lancaster QoS-Architecture extends these interactions into a multimedia computing environment. The TINA-C recommendations, described in section 4.4, differentiate between the QoS interactions of an object and its environment, and also between different object methods.

We extend the idea of interaction points from the TINA-C recommendations and define the complete set of QoS interactions apparent within a DPE. Figure 3 shows this set, which includes all mathematically possible interactions except for the interactions between an object and a method of event, as we do not consider this to be relevant. Each of the points is arbitrarily assigned a letter so that they can be later referenced.

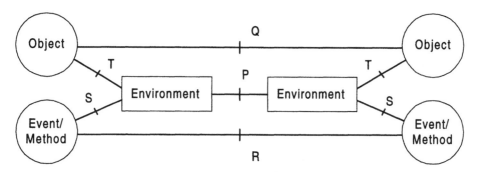

Fig. 3. QoS Interaction Reference Points

Each reference point is considered to potentially require QoS specification. Creating a multimedia service in a DPE involves establishing and negotiating QoS contracts at one or more of these points. The rest of this section details further our defined QoS interaction points and proposes techniques for their specification.

6.1 Reference Point 'P' : *Environmental QoS*

The primary purpose of a DPE is the provision of transparent communication between remote objects. The DPE also provides other services such as object location and user authentication. Requirements of QoS relating to a DPE, concern the quality of these services it provides, that is the services as supplied by the underlying middleware and in particular the remote procedure calling services of the client stubs, skeleton proxies and transport protocols (best classified as non-functional requirements). QoS concerned with these DPE services is termed environmental QoS, and can be mapped onto the QoS requirements of various system components. For example, if a method requests a particular request/reply delay time, this may represent various QoS constraints on methods within a transport object. Figure 4 illustrates this further.

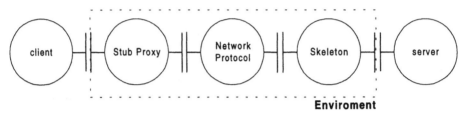

Fig. 4. DPE Environment Objects

Interactions between distributed objects in different domains requires the negotiation of QoS contracts between environments. An example of this is the interaction between ORBs in a CORBA distributed programming environment, where by QoS negotiation of ORB services, such as authentication, delivery guarantee and message security must be made. This allows a client object residing in one domain to be assured the quality of the ORB services in another domain.

Non-functional QoS characteristics applicable to the environment include, cost, message security, invocation accuracy, resilience (the ability to recover from errors), data integrity, authenticability (the ability to supported client authentication), reliability and monitorability (the ability to support QoS feedback from environment).

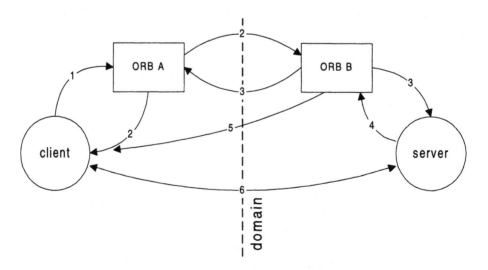

Fig. 5. CORBA Location Interactions

Some functional service characteristics, such as invocation timeliness, can also be defined through static data structures, similar to those as defined by TINA-C. Actual mechanisms for the exchange of environmental QoS is really platform implementation dependent. Specifying functional QoS characteristics with regards to event interfaces, such as for the description of RPC services, are discussed later.

Figure 5 shows the series of messages passed between objects and ORBs within a CORBA environment. Message interactions 2 and 3 are suitable placeholders for the exchange of environmental QoS requirements. One possible approach to facilitating QoS requirements exchange between CORBA environments would be through the extension of the *LocateRequest* and *LocateReply* messages [OMG, 95] as follows;

```
module GIOP {
    struct environment_QoS      {
        short      mean_request_time;
        short      mean_reply_time;
        boolean    authentication;
        short      invocation_accuracy;
        short      security_level;
    };

    enum LocateStatusType {
        UNKNOWN_OBJECT,
        OBJECT_HERE_QOS_SUCCESS,/* indicating QoS agreeable
        OBJECT_HERE_QOS_FAILED, /* QoS not-agreeable
        OBJECT_FORWARD
    };

    struct LocateRequestHeader {
        unsigned long     request_id;
        sequence <octet>  object_key;
        environment_QoS   requiredQoS;
    };

    struct LocateReplyHeader {
        unsigned long     request_id;
        LocateStatusType  locate_status;
        environment_QoS   requiredQoS;
        };
};
```

6.2   Reference Point 'Q' : *Interface-interface QoS*

Interface interactions are concerned with the specification of QoS as provided by the whole set of methods or events. Such specification is important when establishing bindings between objects, where each party must ensure that the other to-be-bound object can maintain certain object QoS contracts. QoS information about an object must be obtainable from both an object interface and an instance of the object since a particular instance may have certain run-time resource limitations imposed on it. A suitable mechanism for the exchange of run-time QoS specification can be done is through methods on a object. Specification on an object interface must is done via the interface definition language itself or some other association.

```
struct camera_QoS {
   boolean monitorability;      /* support for monitoring */
   short cost;                  /* cost of use */
   };

interface Camera : QoS_methods with camera_QoS    {
   void start_stream();         /* camera control methods */
   void stop_stream();
   };

interface QoS_methods   {        /* methods to determine QoS
   short get_remaining_lifetime();          requirements */
   short get_current_loading();     /* ascertain load */
   float get_error_ratio();
   };
```

The methods provided through the QoS_methods interface are used to determine the state of QoS characteristics for a particular object instance during run-time. Such methods may be used by a binding object to determine the suitability of an object before binding to.

6.3   Reference Point 'R' : *Method-Method / Event-Event QoS*

Reference point 'R' defines QoS requirements between single methods or events. A particular event may require a specific QoS from another method in order that it maintain its own QoS contract. An example of this is a multimedia component requiring timing events from a system timer. Operational interface methods may also make demands on other methods.

```
struct method_id {
   char * interface;
   char * method;
   };

struct QoS_control {
   method_id method;
   float availability;
   };

interface Compressor {
   void Stop() with QoS_control;
   };
```

```
interface Control {
  void PostQuitSignal();
  };
```

The method `Compressor.Stop` can only successfully offer its QoS contract provided that the method `Control.PostQuitSignal` maintains a certain availability. Below is an illustration of inter-event QoS requirements described with Lancaster's EDL and a simplified version of QL. The example essentially described the inter-dependencies between a configuration of a camera, compressor and timer objects.

```
events Compressor {
  in FrameReady(RAW_T frame);
  out FrameCompressed(INDEO_T frame);
  out CompressionFailed(MESSAGE_T reason, TIME_T time);
  };

events Camera {
  out FrameCaptured(RAW_T frame);
  out CaptureFailed(MESSAGE_T reason, TIME_T time);
  in SignalCapture(TIME_T time);
  };

events Timer {
  out SignalEvent(TIME_T time);
  };
```

Event QoS Definitions:

```
/* signal every 10 ms */
T(Timer.SignalEvent,n+1) - T(Timer.SignalEvent,n) = 10ms

/* less than 1ms delay between capture and signal) */
T(Camera.FrameCaptured,n) - T(Camera.SignalCapture) < 1ms

/* less than 5ms delay between signal and compression */
T(Compressor.FrameCompressed) - T(Timer.SignalEvent,n) < 5ms

/* mean time between failures */
average( T(CompressionFailed,n+1) - T(CompressionFailed,n) )
> 2000

/* frame dropping ratio */
( ΣCompressionFailed / ΣFrameCompressed ) < 0.001
```

The TORBoyau [Dang, 95] distributed multimedia platform, derived from CORBA, also supports the specification of event interfaces in the form of *signal* interactions.

A QoS clause is optionally associated with each signal as shown in the following example.

```
interface<stream> Telephone    {
   flowIn mike(audio);
   requires QoS(
       "audio_encoding=LINEAR",
       "audio_frequency=22050",
       "audio_precision=16"
       "audio_channels=2");

   flowOut speaker(audio);
   providesQoS(
       "audio_encoding=ULAW",
       "audio_frequency=8000",
       "audio_precision=8",
       "audio_channels=2");
   };
```

The TORBoyau system's extended IDL is also capable of supporting temporal logic by replacing the QoS parameters with language clauses.

6.4   Reference Point 'S' : *Method-Environment/Event-Environment*

Reference point S defines the interaction between a method/event with the supporting distributed programming environment. Requirements, as described in section 5.1, include RPC security, request and reply delay, etc.

6.4.1   Method-Environment QoS

The application programmer may want to associate a particular QoS specification for a single method. An example of such requirement is a method which needs individual costing requirements and whose request/reply time must also be bounded. Performance characteristics which are independent of subsequent calls, such as jitter and synchronisation, have no relevance to one-off operational method calls. Method signatures can attach a QoS definition structure with a suitable place-holder in the interface definition language. One possible technique is appending the `with` clause at the end of the method signature. Below illustrates this approach.

```
interface video_control {
   short get_frame_rate( void ) with monitor_QoS;
   void stop_stream(in short level) with shutdown_QoS;
   };

struct T_monitor_QoS    {
   TIME_T reply_delay;
   };
```

```
struct T_shutdown_QoS   {
  TIME_T request_delay;
  TIME_T reply_delay;
  COST_T cost;
  };
```

Another possible technique, particularly suited to a CORBA environment, is the application of contexts. A context is itself a pseudo-object[6] which represents a set of name, value pairs and simple operations to add, remove and interrogate this information. Contexts are associated with methods in the interface description and provide an ideal mechanism for the exchange of QoS information.

```
interface video_control {
  short get_frame_rate( void )
      context("reply");
  void stop_stream(in short level)
      context("request", "reply", "cost");
  };
```

6.4.2   Event-Environment Interface QoS

In the same way a method can expect a certain QoS from its environment, an event may also make certain QoS requirements (since the only difference between methods and events is their style of interaction). QoS characteristics such as jitter become more important since events usually occur recurrently.

```
struct transport_QoS    {
  float max_request_time;
  };
```

```
events Camera {
  out FrameCaptured(RAW_T frame);
  out CaptureFailed(MESSAGE_T reason, TIME_T time) with
          transport_QoS;
  in SignalCapture(TIME_T time);
  };
```

6.5   Reference Point 'T' : *Interface-Environment QoS*

Finally point T defines interactions between interfaces and their environment. This is similar to reference point S QoS, except that it is concerned with QoS requirements which are specific to the complete set of methods or events, rather than one in particular.

---

[6] A pseudo object is a component which appears to offer methods but in fact is not an object and does not maintain interface reference identifiers.

```
struct camera_QoS {
   boolean authentication;        /* support authentication */
   short security;                /* level of RPC security */
   };

interface Camera with camera_QoS {
   void start_stream();           /* camera control methods */
   void stop_stream();
   };
```

6.6   QoS Specification on Bindings

The term 'binding' refers to the end-to-end connection between one or more service components. It is the binding that provides an agreed end-to-end QoS for the application. Binding QoS is different from the ODP defined 'QoS provided by the binding' which relates purely to the service offered by the transport system and excludes the QoS required and provided by the periphery components. QoS of a binding is achieved through the configuration and negotiation of QoS at the interaction points as previously described.

Binding QoS could be suitably specified during a _bind call. Binding objects, which are a result of a call to a factory object, are service specific and provide a pre-defined service type, e.g. MPEG-1 audio/video stream, CD-audio and Indeo Video. The required QoS for the binding is specific to the type of service. An example of QoS specification on a binding using a CORBA context object is given below.

```
CORBA::Context QoS_ctx = CORBA::Context::Context();

QoS_ctx -> set_one_value("Quality Level",100);
QoS_ctx -> set_one_value("Format",ID_IV32);

binding bo = bind_factory.create("Indeo_Video_Stream");

binding ctrl indeo_control = bo.bind("cam1:camSrv",
                                     "decin1:decSrv",
                                     "decout1:decSrv",
                                     QoS_ctx,&env);
```

Current CORBA implementations of the bind operation do not support QoS specification hooks (mechanisms for associating QoS parameters on a binding), however this is viewed as a simple extension to the environment.

Through the realisation of our previously defined QoS interaction points within a DPE implementation, we can create a set of tools and abstractions ideally suited to the requirements for the provision of multimedia services. The application and DPE then become QoS 'aware', which is a requirement essential to the provision of QoS dependent telecommunication services.

7.   Conclusion

We have discussed the relevant issues concerning QoS specification for multimedia telecommunications services by initially identifying points of interaction which are considered QoS oriented and then applying characteristics from various QoS frameworks to specify QoS at these points. Defining how QoS should actually be specified is a difficult problem to solve since the approach taken depends upon the nature of the service, and therefore it is suggested that only guidelines for QoS specifications within a DPE can be made, as the advent of new services will affect the requirements for specification.

Currently we are using the ideas presented in this paper in the development of a distributed programming environment supporting QoS management. The prototype system will provide the application programmer with the tools necessary to build continuous media stream bindings from a set of user requirements together with the system components for dynamic QoS management. These elements will contribute to a "QoS Manager", which is being developed at Lancaster in a project funded by the BT Labs under their University Research Initiative.

ACKNOWLEDGEMENTS

We would like to acknowledge the kind support of BT Labs in funding this research under the Management of Multi-service Networks project within BT's University Research Initiative (BT-URI). We would also specifically like to acknowledge the collaboration of our colleagues at Imperial College who are also funded under this BT-URI project.

References

[**Blair, 95**] G.Blair and G.Coulson, "Supporting the Real-time Requirements of Continuous Media in Open Distributed Processing", Computer Networks and ISDN Systems 27 p1231-1246, 1995.

[**Bochmann, 96**] G.Bochmann and A.Hafid, "Some Principles for Quality of Service Management", Universite de Montreal, May 1996.

[**Campbell, 94**] A.Campbell, G.Coulson and D. Hutchison, "A Quality of Service Architecture ", ACM Computer Communications Review, Volume 24, Number 2, pp6-27, 1994.

[**Coulson, 95**] G.Coulson, G.S.Blair, "Architectural Principles and Techniques for Distributed Mutlimedia Application Support in Operating Systems", ACM Operating Systems Review, Vol 29, No 4, pp17-24, October 1995.

[Coulson, 96] G.Coulson and D.G.Waddington, "A CORBA Compliant Real-Time Multimedia Platform for Broadband Networks", Lancaster University MPG, June 1996 (to be published in Springer LNCS series).

[Dang, 95] F. Dang Tran, V. Perebaskine, "TORoyau: Architecture et Implementation",
Note Technique NT/PAA/TSA/TLR/4587, CNET, Centre Paris A, 38-40 rue du General Leclerc, Issy-les-Moulineaux, France, December 1995.

[Fedaoui, 95] L.Fedaoui, A.Seneviratne, E.Horlait, "Implementation of End-to-End Quality of Service Management Scheme", Universite Pierre et Marie Currie, QoS Workshop, Paris,1995.

[Halteren, 95] A.Halteren, P.Leydekkers, H.Korte, "Specification and Realisation of Stream Interfaces for the TINA-DPE", Proceedings of TINA'95 Conference, Melbourne Australia, p299-312, February 1995.

[Hensall,88] J.Hensall, S.Shaw, "OSI Explained: End-to-end Computer Communication Standards", ISBN 07458-0253-2 Ellis Horwood, New York, 1988.

[Hutchison, 94] D.Hutchison, G.Coulson, A.Campbell and G.Blair, "Quality of Service Management in Distributed Systems", in Distributed Systems Management (M.Sloman, ed.), Chapter 11,Addison-Wesley, 1994.

[ISO, 95] ISO/IEC, "QoS - Basic Framework - CD Text", Joint ISO/IEC&ITU-T Interim Meeting, Toronto, January 1995.

[ITU, 93] ITU-T Recommendation I.350, "General Aspects of Quality of Service and Network Performance in Digital Networks, Including ISDN",COM XVIII-R 114E, 1994.

[Kawalek, 95] J.Kawalek, "A User Perspective for QoS Management", 3rd International Conference on Intelligence in Broadband Service and Networks (IS&N 95), September 1995.

[Kerherve, 96] B.Kerherve, A.Pons, G.Bochmann, A.Hafid, "Metadata Modelling for Quality of Service Management in Distributed Multimedia Systems", IEEE Multimedia 1996.

[Kitson, 95] B.Kitson, P.Leydekkers, N.Mercouroff, F.Ruano, "TINA Object Definition Language (TINA-ODL) Manual, TR_NM.002_1.3.95, TINA Consortium, 1995.

[Lakas, 96] A.Lakas, G.Blair, A.Chetwynd, "A Formal Approach to the Design of QoS Parameters in Multimedia Systems", Lancaster University (awaiting publishing), 1996.

[Leydekkers, 95] P.Leydekkers, V.Gay, L.Franken, "A Computational and Engineering View on Open Distributed Real-time Multimedia Exchange", Proceedings of NOSSDAV'95 (Springer-Verlag), Boston, USA, April 1995.

[MMCF, 95] Multimedia Communications Form, Inc. "Multimedia Communications Quality of Service", Final MMCF Document , MCFI, 1995.

[Nahrstedt, 95] K.Nahrstedt, A.Hossain, S. Kang, "Probe-based Algorithm for QoS Specification and Adaptation", University of Illinois, QoS Workshop, Paris, 1995.

[ODP, 92] ISO/IEC JTC1/SC21 Draft Recommendation X.903: Basic Reference Model of Open Distributed Processing - Part 3:Prescriptive Model, ANSI, 1430 Broadway, New York, NY 10018, USA, 24th June 1992.

[OMG, 91] "The Common Object Request Broker : Architecture and Specification", Object Management Group, OMG Document Number 91.12.1, http:\\www.omg.org\, December 1995.

[OMG, 95] CORBA 2.0 "Interoperability - Universal Networked Objects", Object Management Group, OMG TC 95-3-xx, 1995.

[Sluman, 92] C.Sluman, "Interoperability Is Not Enough", Communications Week International, Page 20, April 1992.

[Stefani, 93] J.B.Stefani, "Computational Aspects of QoS in an Object Based Distributed Architecture", 3rd International Workshop on Responsive Computer Systems, Lincoln, NH, USA, September 1993.

[TINA, 94] TINA-C, "Overall Concepts and Principles of TINA", TB_MDC.018_1.0_94, February 1995.

[Vogel, 95] A.Vogel, B.Kerherve and G.Bochmann, "Distributed Multimedia and QoS: A Survey", IEEE Journal of Multimedia, Volume 2, Number 2, 1995.

Generic Conversion of Communication Media for Supporting Personal Mobility

Tom Pfeifer, Radu Popescu-Zeletin

Technical University of Berlin
Open Communication Systems (OKS)
Franklinstr. 28, 10587 Berlin, Germany
e-mail: pfeifer@fokus.gmd.de

Abstract: The Intelligent Personal Communication Support System is introduced as an application for multiple media conversion tools, embedded in a context of personal mobility, service personalization and service interoperability support. After discussing models for conversion in theory, the current conversion technology is evaluated. The necessity of an integrated framework of flexible converters and a generic converter model are derived and automatic management of conversion quality is discussed.

Keywords: Media Conversion, Quality of Service, Personal Communication Support, Personal Mobility, Service Personalization, Service Interoperability, TINA Service Architecture.

1. Introduction

A lot of effort has been spent in the previous years to build a world full of multimedia systems. We have experienced great progress in creating digital, computerized multimedia data, compressing and decompressing them efficiently, distributing them over innovative high speed networks, and presenting them with best effort quality on dedicated multimedia devices and workstations.

At their workplace, people can now communicate with each other in high quality, employing all means of communication media their environment supports. However, when they leave this well equipped place, they are not willing to be cut off the information stream. Therefore, they carry mobile communication devices, and they find less properly or differently equipped places at home, in a hotel, or at their alternative workplace.

While they now have to tolerate a lower style of presentation of the information, e.g. because they cannot attend the video conference, or cannot receive faxes or electronic mail with their simple telephone, they are still interested in the semantic of this information. This is the point where the conversion of different communication media can be introduced, with the effect of largely increasing flexibility regarding the choice of communication devices and the transported media. [1]

Currently, such media conversion is done as a dedicated stand-alone process. Faxes received by a computer can be converted to text by the OCR software (Optical Character Recognition), built into the fax software. A speech synthesis program (TTS: Text to Speech) might read incoming electronic mail to the recipient. These processes have to be configured manually. It is difficult to choose an appropriate process automatically, and trying to combine such conversions may lead to unpredictable results, because it is vulnerable to error propagation.

Therefore, this paper investigates a generic approach of converting various media into each other, preserving the semantic of the information contained, complete or in parts. It aims to choose converters automatically for a specific problem in communication, involving the controlled combination (concatenation) of various converters. The conversion should work in distributed computing environments [4, 11].

This approach is applied to the concept of *"Personal Communication Support (PCS)"* which provides people with a new dimension in communication. In general, the concept allows users to establish their own personalized communication environment by addressing three important aspects, namely:

- *Personal Mobility*, which denotes the mobility of the user in *fixed* networks and wireless networks, allowing the user to make use of communication capabilities available at different locations, i.e. at any place;

- *Service Personalization*, including personalized call / reachability management allows the user to configure his communication environment and control his reachability according to his specific individual needs, i.e. *if, when, where, for whom* he will be reachable; and

- *Service Interoperability* in distributed multimedia environments, addressing the interoperability between different types of communication services and terminals, allowing users to maximize their reachability. In this context, the multimedia capabilities are required that enable dynamic content handling and conversion of different media types and formats in order to deliver information in any form.

Therefore the GMD Research Center for Open Communication Systems (FOKUS) and the department for Open Communication Systems at the Technical University Berlin have developed on behalf of DeTeBerkom , a subsidiary of the Deutsche Telekom

$I_1 (M_1, F_1, S_1)$ → media converter → $I_2 (M_2, F_2, S_2)$

Fig. 1.   Media converter system

AG, a TMN-based *Personal Communication Support System (PCSS)* [10] [11], which addresses the first two aspects of PCS, namely personal mobility and service personalization. In order to provide full PCS capabilities within emerging CORBA/TINA based environments, both departments started in 1996 the joint development of an "Intelligent Personal Communication Support System (iPCSS)".

The paper is outlined as follows: Section 2 develops theoretical models for multimedia conversion and introduces the concept of converting media and formats. Section 3 focuses on current converter technology and considers the state of the art of some complex problems. It derives a generic conversion matrix, which quality-of-service management is discussed in Section 4. Then, Section 5 describes the application of conversion in the context of Deutsche Telekom research projects for enhancing Personal Communication Support [18, 19]. After a quick introduction in relevant TINA concepts, chapter 6 describes the iPCSS architecture, its components and their interaction. The remaining sections conclude the paper and provide acronyms and references.

2. Multimedia Conversion

2.1. Modeling

A media converter may be defined as a system entity, which input is information I_1 with the semantic S_1, carried by a specific medium M_1, using a specific form (or format) F_1. We obtain information I_2 as output in another Medium M_2 in format F_2, carrying a semantic S_2 (see Figure 1).

The quality of conversion can be measured by comparing the input and output semantic, S_1 and S_2, which should be preferred to be as close as possible, or having a predefined reduction.

Considering the well-known Shannon's model for information transmission [20], consisting of source and drain, encoding and decoding layers, and the transmission channel, the media conversion could be seen as a specific kind of encoding or decoding, and therefore placed among the encoder/decoder stacks (see Figure 2a).

However, this first approach of modelling fails when the converted information is retransmitted over a different channel after converting, instead of being converted near the sender or recipient. Additionally, the conversion process might require further decoding and encoding of the received and transmitted signal, at least in the physical layer, but often in higher layers as well. Therefore, Figure 2b provides a more general model of conversion. It reminds us of a protocol stack in a network layer model, in particular of the bridge or gateway functionality. Indeed, the conversion process could be considered as bridging two different worlds of communication.

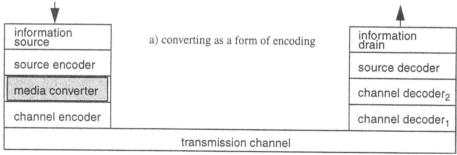

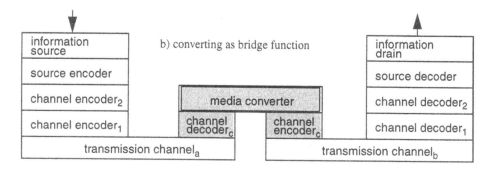

Fig. 2.   Modelling information transmission and media conversion

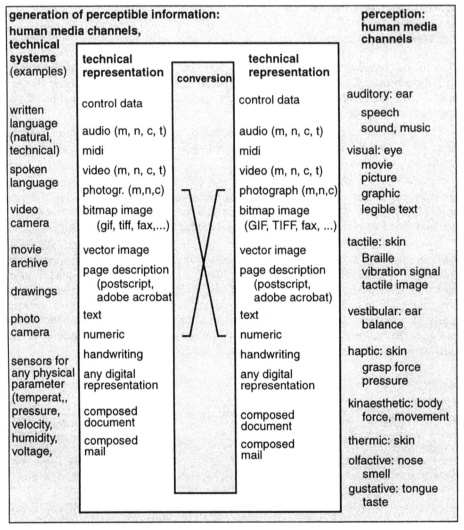

generation of perceptible information: human media channels, technical systems (examples)	technical representation	conversion	technical representation	perception: human media channels
written language (natural, technical)	control data audio (m, n, c, t) midi		control data audio (m, n, c, t) midi	auditory: ear speech sound, music
spoken language	video (m, n, c, t) photogr. (m,n,c)		video (m, n, c, t) photograph (m,n,c)	visual: eye movie picture graphic
video camera	bitmap image (gif, tiff, fax,...)		bitmap image (GIF, TIFF, fax, ...)	legible text
movie archive	vector image		vector image	tactile: skin Braille
drawings	page description (postscript, adobe acrobat		page description (postscript, adobe acrobat)	vibration signal tactile image
photo camera	text numeric		text numeric	vestibular: ear balance
sensors for any physical parameter (temperat,, pressure, velocity, humidity, voltage,	handwriting any digital representation composed document composed mail		handwriting any digital representation composed document composed mail	haptic: skin grasp force pressure kinaesthetic: body force, movement thermic: skin olfactive: nose smell gustative: tongue taste

parameters:
m, n: media dependent parameters
(frame/sampling rate, quantization, resolution, size, color depth, etc.)
c: applied compression technique
t: time, duration, etc.

Fig. 3.   Generic conversion matrix

For specifying the model in more detail it is now necessary to consider the kinds of intended conversions. For this purpose, we need to consider that all information is intended to be perceived by human beings. However, this information could be generated by other humans as well as technical systems.

Figure 3 lists on the left hand side some example sources of information. In the inner block, the technical representations of these information sources are specified. A technical converter would now change one technical representation into another, trying to leave the semantic intact, or to produce a dedicated subset of the input semantic. The technical representation at the output of the conversion can than be applied to the human senses, serving as perception channels for the information.

2.2. Media Type Conversion vs. Media Format Conversion

As we see in Figure 3, even the same class of technical representation can have various parameters, some specific to a certain medium (e.g., colour depth of a picture), some more general (time parameters). Obviously, modifying important parameters of the same medium can be considered as a conversion too.

Therefore, we can distinguish two major classes of conversion:

- Media type conversion, and

- Media format conversion.

The latter class of conversion does not alter the medium type. It converts into another format within the same type, or modifies the medium within the same format (e.g., scaling, colour reduction, etc.). A medium format converter is expected to be highly flexible, allowing most possible conversion combinations, accepting as well as delivering the formats with maximal parameterization (size, sampling rate, etc.).

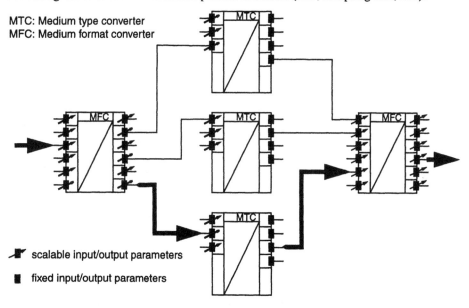

Fig. 4.   Medium type conversion with format adaptation

This first class of conversion alters the type of the medium. As we will see later, the converter is in most cases specialized on a specific task, because it is highly complex (like TTS, OCR, speech recognition, etc.). The converter is expected to be very limited in acceptance of input formats and generation of output formats. Implementing multiple converters, e.g. from different manufactures, in a generic environment, leads to the necessity to accompany the type converter with two format converters at the input and the output, respectively, in order to achieve a larger range of possible formats.

This concept leads to a rudimentary framework of type and format converters, as drafted in Figure 4, where multiple, different implementations of a type converter for a specific task of conversion are accompanied by an input and an output format converter. Together, they form the smallest structure in a converter framework. Note that in case of concatenating two or more type converters (MTC), i.e. two or more structures like Figure 4, the intermediate format converter (MFC) can be utilized by both type converters commonly.

The figure demonstrates the availability of three different type converters, delivering and accepting different input/output formats. For a specific task of conversion, a conversion path has been chosen (fat arrows) which is most appropriate to the current resources and the desires of the user.

Having implemented a working universal conversion matrix as depicted in Figure 3 would enable the system to convert any possible medium into any other. While this would be academically demanding, practical considerations will lead to some discretions that are required within a given environment. These are discussed in the next section.

3. Conversion Technology

3.1. Usability

While the previous section approached the theoretical modelling of media conversion, this one should discuss the technology available, and consider practical constraints.

Table 1 presents a collection of practical examples for illustration, focusing on requirements in multimedia communication and information systems, with the attitude that even conversions that sound strange in the first place might be of very practical relevance. E.g. reading temperature values over a phone line is a conversion of temperature into audio, or converting audio into a vector graphic may visualize the structure of a recorded message.

The chaining of converters is illustrated for the temperature example, which has been implemented for demonstration purposes (Figure 5).

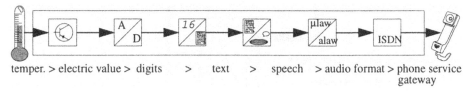

temper. > electric value > digits   >   text   >   speech   > audio format > phone service
gateway

Fig. 5.   Converter chain for temperature to speech conversion with telephone delivery

The range of conversions varies tremendously in effort, cost and required resources. Some kinds are easy to implement with two lines of C code without any patent or license requirements (like converting one audio sample into another), while others are highly complex, requiring sophisticated solutions that might be available only as commercial products, and may be covered by patent or copyright protections (e.g., TTS). Some conversions are purely algorithmic, while other require approaches of artificial intelligence (e.g., speech recognition), or require pipelined processes of decoding, editing, and re-encoding.

source	drain	process	example / application
text	text	format conversions	ASC II 7-bit -> 8-bit
bitmap image	bitmap image		tiff ->JPEG
video	video		MPEG -> H.261
audio	audio		µ-law -> a-law
bitmap image	vector image	vectorization	
audio	vector image	visualization	length and dynamic of a message
text	speech	speech synthesis	TTS
speech (audio)	text	speech analysis & recognition	commands, dictation
bitmap image	text	OCR	OCR
fax bitmap	speech	OCR+TTS	fax reading
temperature	bitmap image	temperature distribution of objects	weather map, medical map
temperature	audio	temp -> text -> TS: value reading	weather report
numerics	image	visualization of statistics	charts
text	tactile information	feed Braille output device	blind reading
control data	tactile information	vibration device	pager signalling
audio	control data	speaker recognition	prioritizing, authorization
photograph or video	text	face recognition, mimic recognition	(e.g. very low bitrate compression)

Table 1   Selected examples of media conversion

A generic intermediate format is often used to reduce the number of required tools in the conversion process. The latter may be of the same kind than the converted medium (e.g. image format conversion), or a different one (e.g. fax > [ocr] > text > [speech synthesis] > reading). In order to avoid losses in quality the intermediate format temporarily needs the resources of the highest possible quality among the involved formats. E.g., in converting bitmap images, the intermediate format needs resources for the largest possible pixel resolution and true-color depth per pixel.

The advantage of requiring less tools is paid with

- the necessity of more resources (e.g. for an intermediate format or for computing time),

- the necessity to convert twice, causing a delay and – in some cases – a reduction in quality.

A possible solution for this problem is a hybrid approach, i.e. to use dedicated tools if available, and intermediate formats in other cases. At this point, it becomes necessary to classify the demands of conversion:

1. highest priority, very often used, maximum speed required:
   This category should be implemented as a dedicated service, featuring a dedicated software solution including supplemental hardware, for performing a one-step conversion.

2. medium priority, often used:
   In the intermediate area compromises are possible, dedicated software would be an option, one-step conversion recommended.

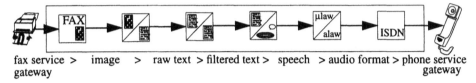

fax service >   image   >   raw text > filtered text >   speech   > audio format > phone service
gateway   gateway

Fig. 6.   Converter chain: fax reception, conversion to text and speech, telephone delivery

3. lower priority, rarely used:
   For services that are seldom employed, the conversion via a generic intermediate format is possible. The resulting multiple-step conversion requires a preliminary definition (or dynamic configuration) of a conversion path, e.g. instructions which converters and intermediate formats are appropriate to achieve the maximum quality or shortest conversion time, respectively.

Even when the same pair of media is converted, different performance, quality and gradability might be required – depending on the application. E.g., speech recognition (speech –> text) can be focused on dedicated speakers for dictation purposes, or on a very limited vocabulary produced by unknown speakers for command or keyword recognition.

3.2. State of the Art in media type conversion

For media conversion, the most complex problems of TTS, OCR and speech recognition should be shortly discussed. All three of them have strong relations to Natural Language Processing/Representation and Artificial Intelligence.

• *Text-to-speech*

Speech synthesis is the task of transforming written input to spoken output. The input can either be provided in a graphemic/orthographic or a phonemic script, depending on its source. Two major classes of algorithms are the concatenation of previously recorded or generated utterances, or the synthesis of waveforms according to a model of the human voice tract. [22, 23]

In the former case, the easiest way is to just record the voice of a person speaking the desired phrases. This is useful if only a restricted number of phrased has to be produced, like pre-specified messages. More sophisticated are algorithms which split the speech into smaller pieces, down to phonemes or their duplications, diphones, as well as syllables. Most commercial TTS systems employ these methods.

The latter case requires format synthesis, which is done by digital signal processing. This version is in more experimental states.

While an intelligible speech can be synthesised today, it sounds still rather artificial and monotone. Therefore, research goes on to apply and improve the prosody of the speech, that is the intonation and phrase melody. While it can partly be done by analysing the grammatical structure of a sentence, it is difficult to find out which words have to be stressed for their relevance to the meaning of the statement. This problem can only be solved applying background knowledge and artificial intelligence. [24] This technique is even more important for the other speech related conversion: automatic speech recognition.

Most manufacturers offer text reading and e-mail vocalization as one of their primary applications. However, in most cases the service is not generic but proprietary, tailored to a dedicated environment of operating system and hardware platform. Mostly, audio is sent to the speaker line without the possibility of further processing, audio format conversion and forwarding.

* *OCR*

Optical Character Recognition identifies valid text blocks in an image, and maps the image representation of the symbols to their ASCII equivalents. Remaining ambiguities in recognition may than be processed by spelling checkers and language processors.

Most systems are designed for recognition of printed characters. Recognition of hand-written language experiences similar problems as speech recognition in recognizing boundaries and requiring more sophisticated interpreting systems.

Many commercial systems perform 99.9% correctness for clean text images in a small number of simple fonts. It is difficult to deal with font and scale variations (omni-font), as well as with noise (from digitizing and intensity variations) [21], particularly when the input comes out of a telefax system, or when included graphics have to be considered.

The recognition of typographical pattern written by hand is a specific sub-class of OCR. It is much more complex due to the variation of people personal handwriting styles, and the wide unlikeness between the writing style of different people.

Therefore, similar to speech recognition discussed below, two major trends lead to useful applications. Firstly, the system could be trained on a specific writer. This approach is very useful for personal or personalized devices, such as handheld computers (e.g. Apple Newton). Secondly, the set of legal characters for recognition could be limited (similar to command recognition in speech). This approach is widely used e.g. in sorting letters automatically in snail-mail distribution centres.

* *Speech recognition*

Automatic speech recognition is the process by which a computer maps an acoustic speech signal to text. Automatic speech understanding is the process by which a computer maps an acoustic speech signal to some form of abstract meaning of the speech.

Such systems can be speaker dependent, adaptive, or independent. The first kind is developed to operate for a single speaker, therefore usually easier to develop, cheaper to buy and more accurate, but not as flexible as speaker adaptive or speaker independent systems. A speaker independent system is developed to operate for any speaker of a particular type (e.g. American English). These systems are the most difficult to develop and most expensive to buy and their accuracy is lower than speaker dependent systems. However, they are more flexible. [25, 27]

The size of vocabulary of a speech recognition system affects the complexity, processing requirements and the accuracy of the system. Some applications only require a few words (e.g. numbers or commands only), others require very large dictionaries (e.g. dictation machines).

Another problem relates to continuous speech vs. isolated-words. In the latter case, a pause between saying of each word makes the recognition of word boundaries easier. In continuous speech, there are much less hints for these boundaries, additionally, neighboured words influence each other (coarticulation).

The process of speech recognition starts with the digital sampling of speech, followed by acoustic signal processing, mostly including spectral analysis. Next, phonemes, groups of phonemes and words have to be recognised. Systems based on Hidden Markov [26] modelling are mostly used.

3.3. State of the Art in media format conversion

For converting one format of the same medium into another, tools exist for many platforms for text, bitmap images and audio. They are mostly in the public domain, and perform well as software solutions [32]. Converting video formats requires the appropriate encoding/decoding hardware and software for the compression methods involved [28, 29, 30, 31].

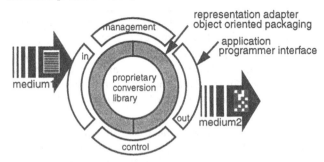

Fig. 7.   Generic converter model

3.4. Generic Converter Model and Generic Converter Framework

Like outlined above, a *generic converter model* has been designed. A generic converter framework has been implemented according to the converter model in order to abstract from the requirements of the underlying converter software.

In the generic converter model shown in Figure 7, the conversion is executed in a core library, which is proprietary to the manufacturer. A representation adapter layer covers the specific properties of the core library and unifies the format requirements for the conversion framework. For extrinsic converter solutions not complying completely with the object oriented design of the framework, the representation adapter also realizes the object oriented behaviour.

The outer layer delivers management and control interfaces. Input and output adapters modify the behaviour of the converter to specific requirements of the medium type and might perform some pre- or post-processing.

In order to keep track with personal communication related research activities and the corresponding platform developments, the iPCSS will migrate towards conformity with the Telecommunications Information Networking Architecture (TINA), developed by a consortium which is currently in the focus of worldwide attention [13]. This architecture is based on distributed computing (e.g. based on CORBA [16]), object-orientation, and other concepts and standards from telecommunications and computing industries.

The details of the converter model are specified in this project in a way that makes the migration towards TINA as easy as possible. [10]

In a framework, particularly in a distributed environment [16], various tools of conversion can be provided for different requirements, with different performance, and at different cost. To form such a framework, the conversion tools utilized must have unified interfaces, the *Generic Converter Interface* (GCI) for various purposes.

The GCI defines initialization, the localization of an appropriate converter, access to the description of conversion quality (see next section), data flow control, management, and finally termination of an instance of a single converter.

The *generic converter framework* has to abstract from the underlying hard- or software used for the actual conversion. E.g. for first testing scenarios, widely used software packages for simple format conversion and complex solutions for OCR, TTS and FAX have to be included by simplifying the access to these packages as much as possible.

The main purpose of the converter framework is to identify the general behaviour of filter and converter software packages and to define a unique subset of functions which can control, start, stop, manage and configure these software packages. Furthermore, general methods for the deliverance and transport of the conversion data have to be identified.

Currently used software packages can read and write there data from files, pipes or even network-wide sockets. The implementation of the converter framework realizes the GCI and hides the specific maintenance and data transport functions from calling objects like SAPs. Additionally it maintains the QoS matrix (cf. section 4.) of all available QoS parameters for all software packages.

The converter framework initiates and reserves conversion services, validates the state of the started hard- or software, controls jobs and processes as well as the data flow via file redirection, pipes or TCP/IP-based services and database functions for the specification of available soft- and hardware modules used for the conversion.

4. Quality of conversion services

As it was stated above, conversions are of different quality due to the nature of the processes involved, and due to limitations in current technology. The conversions might be lossy (like or because of lossy compression), quality-limiting (like conversion to lower resolution), or error-prone (like OCR or speech recognition). When such conversions are applied, these quality problems have to be considered. [33]

4.1. Quality control and management

Each conversion consumes resources of the environment and influences the quality of the final output. When concatenated, the influence of each conversion has to be considered in the context of the whole chain of converters. Objective criteria are required for choosing the optimal conversion path among various possibilities for a specific task. However, different tasks can emphasize different parameters for the selection. E.g., synchronous communication requires a short over-all delay, while asynchronous conversion (for later delivery) might emphasize minimal cost or best intelligibility of the output.

4.2. Quality-of-Service definition

Similar to the Quality of Service (QoS) parameters in network communication, a lot of parameters could be defined for conversion, encompassing:

- *Intelligibility*
  This parameter is the most important determiner for the correct transport of the semantic of the information during the conversion process. Defined in Webster's dictionary as "capability of being understood or comprehended" or "apprehensibility by the intellect", it describes whether the human perceiving the output of the decoding process is able to recognise its semantic correctly. The term is mostly used in the context of complex conversion, as TTS, OCR and speech recognition.

- *Error probability*
  This parameter is the more technical version of the previous. It describes the probability of bit errors both in conversion and in required transmission between conversions. Both parameters, the intelligibility and the error probability need to be considered in their interference.

- *Quality degradation* due to *lossy compression/decompression*
  This parameter should prohibit multiple lossy compression and decompression processes, if possible (e.g., an image should be compressed into JPEG [31] format only finally, not as an intermediate step).

- *Quality degradation* due to *entropy reduction*
  This parameter describes the loss in quality (and therefore, finally, in intelligibility) when the semantic of the information is partly reduced (e.g. because a 24 bit colour image has to be displayed on an 8 bit screen, i.e. colour reduction, quantification, scaling).

- *Delay*
  Depending on the kind of conversion, there are a start-up delay for a continuous stream of data, delay caused by buffering processes (e.g. an image might be buffered completely before conversion, or a TTS software might wait for the end of a sentence or paragraph before determining the correct prosody), and transmission delay. Such irregularities in converters can produce **jitter** to an considerable extent.

- input/output *data volume and data rate*
  This parameter describes the required storage and transmission resources.

- *Computational resources*
  This parameter determines the requirements of the conversion process regarding hardware platform and computational performance. It has to be considered that a single conversion might be possible on a specific service computer, e.g. speech synthesis, while for multiple parallel conversions specific hardware (like a board of parallel signal processors) might be required.

- *Cost* (respecting tariffs)
  This parameter refers to the computational resources above as well as transmission cost, but is calculated for a single conversion process.

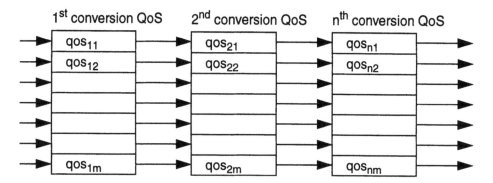

Fig. 8.   Propagation of QoS-parameters in concatenated conversions

Comparing different possibilities of concatenating converters for a specific task requires a complex evaluation of the quality parameters involved, performed at runtime. Figure 8 illustrates the propagation of the parameters in concatenated conversions.

For each parameter, the dedicated model for concatenation has to be derived. While delays probably add up, quality degradation and error propagation require more sophisticated calculations, like the evaluation of the sum of squared differences, etc.

4.3. Media dependence of QoS parameters

Depending on the existence and the character of losses in the conversion, the process can be reversible or not. E.g., bitmap image format conversion is bitwise reversible as long as color space and resolution are not touched. On the other hand, even printing the output of OCR back on paper will suffer from font modification and formatting losses.

Therefore, this sections emphasizes the difficulties to compare QoS parameters of completely different media due to their heterogenous characteristics. For example,

- *Text* may suffer from character reduction, loss of layout formatting,

- *Images* can be reduced in color space, size and compressed lossy,

- *Audio* can be reduced in sample size, sample rate, suffers from compression loss, and is specifically sensitive to jitter.

- For *Video*, the huge amount of raw data requires compression (complex but effective) in most cases. The standardization efforts have led to a few platform independent formats, however, the required re-encoding in a conversion causes a considerable delay. The problems of still images apply here also, and the unique parameter framerate influences the acceptance by humans severely.

4.4. Evaluation and handling of QoS

Algorithms for evaluating various choices of concatenated conversions for the purpose of automatic configuration are currently under development. Most promising are backtracking strategies for analyzing the multiple possibilities, and fuzzy logic comparison of the heterogenous QoS parameters.

The latter method has the advantage that no hard limits have to be defined for individual parameters. Instead, weak parameters can be compensated by other advantages of a chain configuration. E.g., a slightly too long delay might be accepted if the service is cheap enough.

In a first approach, a selection of four important QoS parameters has been made, which cover most considered aspects above. These parameters are:

- *Intelligibility*, covering problems of media synthesis, error probabilities, compression losses, channel noise, and semantic reductions, given in percent (0...100%).

- *Bandwidth*, covering data volume and bitrate, given in bit/s,

- *Delay*, given in seconds

- *Cost*, covering all aspects of tariffing and resource consumtion (transmission and computation), given in countable units.

The narrowest bandwidth of all used components limits the bandwidth of a converter chain. Delay and cost add up and must not exceed predefined limits (depending on the importance of the message). Intelligibility percentages of concatenated components can be multiplied, so that the result is smaller than each value. More sophisticated evaluation models are subject of current research.

Furthermore, the iPCSS is controlled by its users. Consequently, measures taken to cope with the quality problem have to be under their control as well. Such measures could include:

- *Carbon copies:*
  The system keeps copies of the original message, even if it is converted to another medium or another format, and forwarded to another location. For example, faxes that are converted to text (OCR) are kept as the original bitmap for later proving.

- *Limitation of concatenation:*
  Knowing the artifacts of the conversion tool involved, the user might want to limit concatenated conversion steps. E.g., if the OCR produces text results that are only legible with background knowledge, then sending them through a TTS system would make things worse and produce an incomprehensible output. In such a case, concatenation would be disabled after the OCR. Additionally, recursive or reverse conversion has to be avoided.

- *Media Selection:*
  The user might like to decide which transport or displaying media type he prefers in order to maintain a desired quality (for example costs). A possible selection can range from transport, format or media types the user does not like to be included in the conversion process, as well as a description of the resulting format and media type conversion.

5. Intelligent Personal Communication Support

The vision for future communication is labelled by the slogan *"information any time, any place, in any form"*. This vision is based on the society's increasing demand for "universal connectivity" and the technological progress in the areas of mobile computing and telecommunication.

In the remaining sections, the application of automatically configured conversion technology for Personal Communication Support is discussed.

As outlined in the introduction, the iPCSS (Intelligent Personal Communication Support System) aims for the provision of full PCS capabilities. The definition of the PCS concept is strongly influenced by recent research activities in the field of advanced telecommunication and distributed computer sciences, such as IN, UPT, Telecommunication Management Network (TMN) [8], TINA, computer telephony integration [2], mobile/ubiquitous computing [3], and Electronic Location Systems [9].

The trends toward PCS can be viewed in terms of three major areas of research:

- *Mobility support in fixed and wireless networks,*
  enabled by means of *terminal mobility* and *user mobility. Terminal Mobility* by wireless network interfaces and protocols (i.e. cordless, cellular and satellite) is fundamental for the provision of ubiquitous, global connectivity. The next step, *User Mobility* (also called "Personal Mobility" or "Service Mobility",) will enhance global service access, allowing people to make use of any kind of terminal located at their whereabouts for obtaining access to their services. With this type of mobility support, the user is directly addressed by the means of a personal number instead of addressing the terminal at his guessed current location.

- *Personalization of communication services access and delivery,*
  enabling the user to define his own environment and service working conditions in accordance with his own needs and preferences, with respect to parameters, such as time, space, medium, cost, integrity, security, quality, accessibility and privacy. These parameters are usually stored in a user profile, defining all services to which the user has access, the way in which service features are used, etc. A different aspect of personalization is concerned with the individual management of the user's reachability (when, where, for whom, by what media), i.e. negative/positive communication filtering.

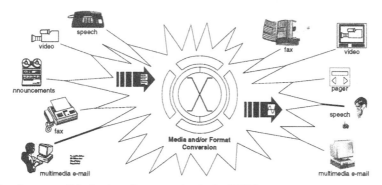

Fig. 9.   Priorized media conversion in the iPCSS

- *Dynamic adaptability, service interoperability and flexible media handling,*
  which is a a much less common aspect of personal communication. *Dynamic adaptivness* means that the communication environment is forced to adjust to the user's needs, knowledge, and preferences as well as to the constraints arising from the user's current local environment. The general intend is that the system adapts to the user and not vice versa. In particular, this adaptation includes flexible media handling in form of possible *conversion* of one medium into another one (cf. Figure 9).

Consequently, the following design criteria for a system/platform providing PCS capabilities can be derived:

- The addressing of users has to be decoupled and made independent from service, network and terminal capabilities. This leads to the *introduction of personal names/ numbers* for achieving real person-oriented communication.

- The user's service specific (access) control data distributed across multiple communication systems and maintained in service specific data structures has to be unified and integrated into a *common user service data structure* (e.g. a generic *"user profile"* or *"personal call logic"*), configured by powerful *customer (profile) management capabilities*. This includes advanced user registration capabilities for personal mobility support, allowing manual and automatic registration at *locations*, such as offices, instead of specific terminals.

- Furthermore, the vision of delivering information in any form requires the introduction of additional concepts, allowing the *dynamic selection of terminal equipment* at a registered location in accord to the incoming service requirements and/or the called user's preferences. In case of inadequate terminal capabilities *service interworking* and/or *media conversion* require the provision of generic service gateways and/or media converters, as discussed in the previous sections.

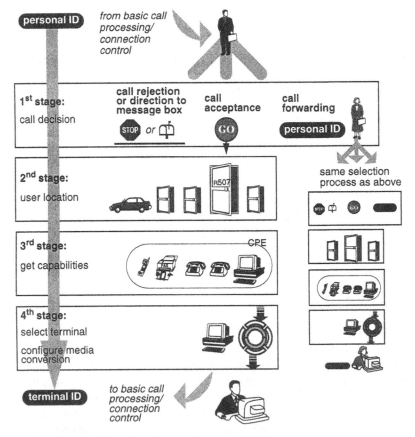

Fig. 10.   PCS-based Intelligent Call Processing

Taking all these issues into account the concept of PCS aims for person-oriented and location-oriented communication. Thus PCS enhances the reachability of persons on the one hand while providing reachability control on an individual basis on the other hand. Note that the PCS capabilities are generic, i.e. these capabilities should be provided by an open service / communication platform via a generic application programming interface to many communication services in a uniform way.

In order to illustrate the benefits of the PCS concept, Figure 10 depicts a simplified intelligent call processing model to be performed by an advanced Personal Communication platform. It is characterized by a *four-stage mapping process* that translates a logical user name used as the called party address (i.e. a personal ID) into an appropriate network address (i.e. a terminal ID). This temporary physical address is passed back to the requesting communication service. The mapping process looks as follows:

- 1st, the evaluation of a user's "Personal Call Logic" provides the *control of his reachability*. The result may be a forwarding to another user, a call rejection, a call redirection to an asynchronous service, e.g. an answering machine, or an acception.

- 2nd, the exact recipient of the communication invitation has been settled and no further call management will be performed. A *mapping of the user to his location* is made based on user registration data.

- 3rd, it *maps a location to a virtual communication endpoint* corresponding to a terminal group representing the set of all access devices in the user's current vicinity. An object-oriented modelling of virtual communication endpoints encompasses the knowledge on terminal capabilities, supported services, and selection mechanisms.

- 4th, an *appropriate terminal ID* from the group of devices is selected and parameterized by a service type, used communication media, and optionally by user preferences. Within this stage, two cases can be distinguished:
  a) In case there exist at least one device of the virtual communication end-point supporting the desired medium of the call, the most appropriate device is selected.
  b) In case no device for the desired medium can be found, further rules of the Personal Call Logic determine whether a *conversion into another medium* is allowed/restricted. Then, the necessary converters are configured and a now appropriate device is selected.

6. The iPCSS Architecture – compliant to the TINA Service Architecture

It is beyond the scope of this paper to introduce the whole TINA concept behind the iPCSS. However, some terms should be introduced for common understanding.

6.1. Basic TINA concepts of an Access Session

The overall TINA architecture is divided into several architectural aspects [13]. However, within this paper we will focus on the TINA Service Architecture [14]. The latter defines a set of concepts and principles for the uniform design, specification, implementation, and management of telecommunication services and their components. It provides the universal platform for a variety of services in a multi-provider environment. This section concentrates on the most important aspects of the Service Architecture, namely the *"session concept"*, which separates different aspects like service access, service provision and service communication needs, and the generic *"service components"* identified within the service architecture. It has to be stressed, that this description is not complete and highlights only the basic concepts in order to understand the iPCSS design.

Since TINA is intended to support even complex multimedia, multi-party telecommunication services, TINA introduced the notion of a "session", replacing the traditional notion of a "call". It provides a means for grouping specific activities in a service during a specific period of time. Three basic types of sessions have been defined [12] [14]:

- The *Access Session* supports an user in accessing, requesting and retrieving telecommunication or information services. The Access Session is service independent and dedicated to a user.

- A *Service Session* represents the core functionality of a service and is therefore service specific. It provides a user or a group of users with an environment to support the execution of a service.

- A *Communication Session* provides an abstract view of connection related resources and supports the activities needed to establish the communication channels (i.e. streams) that may be required between end user systems (and the service provider systems).

TINA services are described in terms of interacting components in distributed processing environments, i.e. *Computational Objects (COs)*. In accord with the session types outlined above, the TINA service architecture introduces a set of generic components for the realization of telecommunication services. This means that each session type is realized by a specific set of interacting COs. In the following we address the main COs defined within the Access Session, which are service independent, and some generic Service Session COs. Note that all these COs can be considered as the generic "construction kit" for the realization of TINA-based telecommunication services.

The generic COs of a TINA Access Session are shortly introduced here, used within both, the user's own end-system domain and the provider domain.

The customer's end-system is represented by two COs:

- The *User Application (UAP)* models the specific service application in the user system, i.e. it offers the application's user interface. It provides the user with access to the Service Session. The latter is modelled by a Service Session Manager (SSM) CO and multiple User (Service) Session Manager (USM) COs, created dynamically by a Service Factory (SF) CO within the provider domain.

- The service independent *Generic Session End-Point (GSEP)* models the minimal set of capabilities required for interacting with the User Agent (see next) to perform service session control and invitation delivery.

Within the provider's domain, the following COs relate to the Access Session:

- The *User Agent (UA)* represents a user in the service provider domain. It is the contact point of control for personalized session creation, suspension, resumption and deletion.

- The *Personal Profile (PPrf)* maintains the user related constraints and preferences on service access and session execution. It determines the environment, in which the service will be executed for the user.

- The *Usage Context (UCxt)* maintains the knowledge on a pool of resources available to the user for the execution of services. It contains registrations at user terminals and terminals at network access points. For personal mobility support, the UCxt keeps track of the terminals and access points available to the user.

- The *Terminal Equipment Agent (TE-A)* represents a terminal of a user system within the service provider domain. It maintains the capabilities and the state of a terminal from the providers perspective.

- The *Subscription Agent (SubAgt)* is a contact point for accessing subscription information according to users. It interacts with other management related computational objects beyond the scope of thispaper.

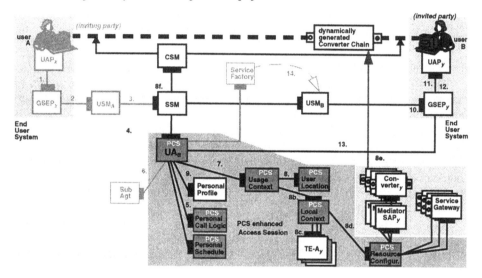

Fig. 11.   Components of the PCS-enhanced TINA Access Session

6.2. The iPCSS Architecture

The iPCSS architecture is strongly based on the TINA Service Architecture. In general, the capabilities of the iPCSS aim for an enhancement of the TINA Access Session in order to increase the nomadic user's reachability by introducing location-based user registration and coincidentally the dynamic selection of terminals at a registered location. In particular the iPCSS allows the adaptation of information flows to a user's terminal capabilities. Therefore, the iPCSS architecture, which could be considered as an enhanced TINA Service Architecture, defines a set of new COs, mostly related to the Access Session, in order to achieve the intended functionality. In the following we look more detailed at these COs and their interactions.

6.3. iPCSS Components

The Access Session enhanced by iPCSS components is given in Figure 11. The following COs have been redefined or newly defined:

User Agent

The *PCS-enhanced User Agent (PCS-UA)* is a specialization of the TINA UA defined in the TINA Service Architecture [13]. The PCS-UA has been designed according to the following requirements which are in line with the TINA Service Component Specifications [15]. It

- controls *incoming calls / invitations to join a session*, i.e. the ability to access a user (i.e. alert, instantiate an USM on behalf of the user) and thus protects the user from unwanted communication attempts (cf. Personal Call Logic). Therefore it can be customized (personalized);

- provides (personal) mobility support to locate and reach a user who may move (cf. usage context, user location, local context);

- provides enhanced functionality for inter PCS-UA communication, e. g. for advanced call forwarding which delegates a service invitation to a PCS-UA of the party it is forwarded to.

Personal Call Logic

The *Personal Call Logic* of a user determines, how a service invitation should be handled by the invited party. It is the main component to optimize or limit a users reachability and to personalize incoming call handling.

Firstly, the PCS call logic component decides on behalf of the user if an invitation should be accepted or not. Therefore, it contains a rule based system defining the user's intends how to handle incoming calls (e.g., accept, reject, forward, call screening, etc.). A user may define his personal call handling, such as accepting/blocking/ forwarding service invitations with respect to the caller, the calling time, the service related to the call, etc.

Secondly, the rules determine whether conversions of communication media are allowed if necessary, enforced in a specific situation, or forbidden in certain cases. These pre-evaluated rules are forwarded to the Usage Context for final selection of terminal devices.

User Location

Regarding registration of users, the TINA Service Architecture is designed to handle terminal registrations with the usage context computational object. The more generic PCS approach supports the registration at locations. That moves the terminal selection to the invitation phase. This concept offers the possibility to realize a flexible call handling, that may take current service requirements and network and terminal states into consideration. The possibility of location-oriented user registrations will be achieved by the new *User Location* CO, that supports the PCS-UA by maintaining location information of a mobile user in form of location identifiers.

Personal Schedule

This object serves for user registrations triggered by a personal schedule or diary. The registrations may be terminal registrations or registrations at locations (cf. User Location). In addition, the Personal Schedule may contain data and logic for time-dependent service invocations on behalf of the user.

Local Context

Many of the information to be handled by PCS capabilities are location related. At the current stage, TINA does not deal with location information yet. Therefore, a new component named *Local Context (LCxt)* has been introduced. Like the Usage Context, the LCxt will maintain knowledge on a pool of resources. *Note that the LCxt is not a user specific component, since it models common infrastructure!* In a service invitation scenario, the LCxt will enhance the access session functionalities to resolve a location identifier of a user to an appropriate network access point and terminal ID. Thus, the LCxt has to contain associations between a specific location (e.g. a room or zone) and terminals (in TINA represented by the TE-A).

From the view of the PCS concept, the Local Context models a virtual communication endpoint. The LCxt is designed to support the Access Session implementing the selection of a terminal at its network access point at the user's current location. Potential client objects of the LCxt are the Usage Context and User Location. At this projects stage, the Usage Context is assumed to query the User Location and subsequently the LCxt pertaining to the user's current location to get terminal information appropriate to the requirements imposed by a specific service invitation.

Usage Context

The *Usage Context* represents the communication capabilities available to a specific user. It keeps track of the set of terminals the user has registered. The Usage Context has been extended to provide a reference to the current location of the user (cf. User Location) and the associated Local Context.

The Usage Context selects a terminal between the set of registered terminals according to requirements of the inviting service. In case no appropriate terminal registration is available, the location registration information produced with the help of the User Location CO and the LCxt CO will be retrieved. The Usage Context maintains the dynamically changing association to both former components and decides when to access both.

Terminal Equipment Agent

The Terminal Equipment Agent contains information about capabilities and current state of a terminal, i.e. a physical device able to handle certain media in a specific way. In its specific form within this project (SAP, Service Access Point), the Terminal Equipment Agent is extended with additional interfaces, enabling the dynamic inquiry of properties of represented terminals.

Resource Configurator

The Resource Configurator is the coordinating instance for selection, configuration, and composition of all terminals, media converter functionalities (represented as Mediator Unit SAPs, see below), Service Gateways, etc. assigned to a certain location. In particular, this component is able to configure chains of multiple converters, to evaluate their Quality of Service, and to select the chain most appropriate for the desired task (see Fig. 5). The result of the configuration is a dynamically generated converter chain with stream interfaces for the appropriate media.

Mediator Unit SAP

This component represents the functionality of media converters (described below), based on uniform definitions of a Converter Framework. Each converter is subordinated its specific MSAP. It has been designed for the purpose of dynamic binding of converters. The MSAP provides the knowledge about individual properties of its converter, regarding purpose, range of configurability, parametrization, and its QoS parameters. In case the converter itself is not conform to TINA/CORBA, the MSAP can be used as a wrapper to provide such an interface.

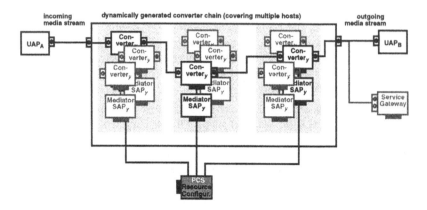

Fig. 12.   Converter chain, configured for a specific task

Medium Type and Medium Format Converters

The design of all Media Converters and Format Converters is based on the open Converter Framework. The various converters can be distributed in a local environment, or even offered as remote services. This distribution is hidden by the Resource Configurator.

Generally, all converters have two types of interface:

- an operational interface towards the Resource Configurator (through the MSAP),

- one or more stream interfaces for incoming and outgoing information flows.

These interfaces are being defined as a bundled Generic Converter Interface (GCI), providing a uniform and flexible access to any Media Converter or Format Converter. The foreseeing definition of this interface is a major key for achieving flexibility to integrate more converters in future developments. The core conversion libraries are adaptations of products which are available commercially or in the public domain.

Specific examples for converters for first implementation are:

- Audio format converters

- Image format converters

- Video format converters

- Text format converters

- Text-to-Speech

- Optical character recognition

- Speech recognition

The complexity of these conversions is highly different, as discussed in the previous sections.

Message Store

This internal service component (cf. Figure 13) provides storage and provision of bulky data of multimedia messages. Such data can be delivered from the *phone.in* service gateway, or from recording tools for customized messages.

Service Gateways

Service Gateways are tools for realizing Service Interworking. They are responsible for transporting information into and out of the context of the iPCSS, i.e. connecting the iPCSS to the world outside the TINA platform. They have to consider the specific properties of the connected information and communication services. In particular, they have to adapt the signalling and digitizing of legacy telecommunication systems.

The connections might be synchronous or asynchronous, and it might be unidirectional (simplex) or bidirectional (duplex).

Specific Service Gateways are:

- *phone.in, phone.out*, and *phone.duplex* for voice connection with the public telephone network; used for recording voice-box messages, delivering prepared audio data, or synchronously connecting phone talks, respectively,

- *fax.out* and *fax.in* for sending and reception of G3 faxes,

- *mail.in* and *mail.out* for reception and delivery of multimedia e-mail,

- *paging.out* and *sms.out* for delivering signalling information to pagers and short message services,

- *mmc*, for support of multimedia conferencing.

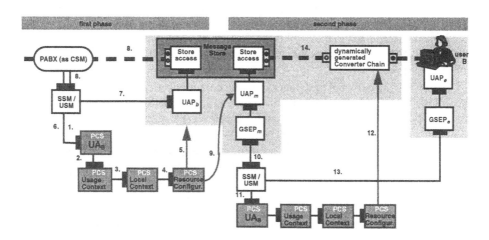

Fig. 13.   PCS-enhanced TINA Access Session for an asynchronous communication example

6.4. iPCSS Component Interactions

- *Registration*

In contrast to the registration at terminals, the iPCSS enables user to register with locations, e.g. offices, in many different ways, mainly manual registration, scheduled registration and automatic registration. The first two registration types can be realized via a specialized *"User Registration Application"*, representing an UAP CO which interacts with the PCS-UA, which in turn interacts with the User Location CO and/or Personal Schedule CO.

In case of automatic registration, i.e. if the user wears an Active Badge [9], no explicit user action for the registration is required, since an *"Electronic Location System"*, modelled as a specific PCS support application will directly manipulate the user registration data in the User Location CO, i.e. modify the references to a corresponding Local Context.

• *Synchronous communication*

Figure 11 depicts the CO interactions within a PCS-enhanced Access Session. An invitation of user B is delegated via user A's UAP_x, $GSEP_x$, USM_A and the SSM to the user B's PCS-UA (1, 2, 3, 4). Within the enhanced Access Session the PCS-UA$_B$ checks first the Personal Call Logic of user B for the reaction to that invitation (5). In the case that the Personal Call Logic indicates to accept the call, the subscription information of user B for the requested service is checked by the respective SubAgt (6). With service information received from the SubAgt, the configuration available for the user has to be retrieved (7). The current location of user B has to be retrieved by querying the User Location (8).

Using the location information, the LCxt attached to that location can be found (8b). The LCxt component allows terminal equipment suited to the service to be selected (8c). In case there is no terminal available supporting the requested media type, the Resource Configurator (8d) is invoked. Considering the knowledge of the MSAPs about the subordinated converters, it selects and configures one or multiple converters dynamically to an appropriate converter chain (for more details cf. Figure 12). The latter is instantiated as an object with stream interfaces (8e). Its reference is then passed back to the SSM (via LCxt, UCxt, UA). The subordinated Connection Session Manager (8f) uses this reference for the connection of the streams to the converter object.

Considering the personal preferences (9), the communication request is indicated to user B (10, 11). If user B accepts the invitation (12, 13), the SF is instructed to create the corresponding USM_B (14). The subsequent service processing establishes the stream connection, with respect to the possible converter chain.

While the previous description fits most for synchronous communication, e.g. a telephone user participating a multimedia conference, an example for the asynchronous case is given below.

• *Asynchronous communication*

The asynchronous illustration (Figure 13) is explained for the example of an incoming phone call, which stores its message in a voice box due to unreachability of the invited user. As in the examples before, the primary invitation is passed via the PABX and the SSM to user B's UA (1). Her personal logic indicates unreachability, and points to the message store (2, 3, 4). The respective Resource Configurator initilizes the User Application of the Message Store (5), and passes its reference back to the SSM (6). The latter (which is for simplification not separated from the USM) establishes the connection to the UAP (7) and connects the stream to the Message Store interface (8). The store works as voice box and records the message, then the call is terminated.

However, the Resource Configurator (RC) has learned the preference of the invited user to get the voice messages forwarded as multimedia e-mail, either in audio form, or even converted to text by speech recognition. In any of these cases, format or media conversion is required.

An important task of the RC is therefore to *trigger* the User Application in the Message Store responsible for forwarding the messages (9). This UAP begins to establish a new call situation, employing the respective SSM/USM (10), and user B's UA again (11). But now the RC creates an appropriate converter chain for the desired task (12), so that the SSM can establish the connection towards the user's Mail-UAP (13, 14), or, alternatively, a service gateway for forwarding a mail anywhere else.

7. Conclusions

This paper has presented an overview of the iPCSS, representing a CORBA/TINA-based platform for the provision of full PCS capabilities as an usage example of automatically configurable technology of media conversion. This technology has been evaluated from the theoretical viewpoint, and quality aspects for the automatic process have been considerd.

As shown in the paper the iPCSS provides enhanced reachability of users while the users are able to control/manage their reachability. PCS capabilities are service generic and are mostly related to the user's access to services.

A prototype implementation of the presented platform will be available by the end of 1996. Further information about the iPCSS can be obtained from "http://www.fokus.gmd.de/ice/".

8. Acronyms

CO	Computational Object
CPE	Customer Premises Equipment
CSM	Connection Session Manager
GSI	Generic Converter Interface
GSEP	Generic Session Endpoint
IN	Intelligent Network
iPCSS	intelligent Personal Communication Support System
LCxt	Local Context
MSAP	Mediator Service Access Point
OCR	Optical Character Recognition
PABX	Private Automatic Branch Exchange
PCS	Personal Communication Support
POTS	Plain Old Telephone Service
PPrf	Personal Profile
SSM	Service Session Manager
SubAgt	Subscription Agent
TE-A	Terminal Equipment Agent
TINA	Telecommunic. Information Networking Architecture
TMN	Telecommunication Management Network
TTS	Text-To-Speech conversion
UA	User Agent
UAP	(End) User Application
UCxt	Usage Context
UPT	Universal Personal Telecommunication
USM	User (Service) Session Manager

9. References

[1] Schmandt, Chris: Multimedia Nomadic Services on Today's Hardware. - in IEEE Network, Sept./Oct. 1994, pp. 12-21
[2] IEEE Communications Magazine, Special Issue on Computer Telephony, Vol. 34 No. 4, April 1996
[3] Weiser, M: Some Computer Science Issues in Ubiquitous Computing. - in: Communications of the ACM, Vol. 6, No. 7, July 1993

[4] Guntermann, M.; et al.: Integration of Advanced Communication Services in the Personal Services Communication Space - A Realisation Study. - in: Proc. of the RACE International Conference on Intelligence in Broadband Service and Networks (IS&N) 1993 (Mobilise), pp. II/1/p.1-12

[5] ITU-T Draft Recommendation F.851: Universal Personal Telecommunications - Service Principles and Operational Provision. November 1991

[6] ITU-T Recommendations Q.121x series: Intelligent Network Capability Set 1. Geneva, 1995

[7] Dupuy, F; Nilsson, G; Inoue, Y: The TINA Consortium: Towards Networking Telecommunications Information Services. - in: Proc. XV. Intern. Switching Sympos., ISS '95, Berlin, Germany, April 1995

[8] ITU-T Recommend. M.3010: Principles of a Telecommunications Management Network. Geneva, 1992

[9] Harter, A; Hopper, A: A Distributed Location System for the Active Office, - in: IEEE Network, Special Issue on Distributed Applications for Telecommunications, January 1994

[10] Eckardt, T; Magedanz, T; Popescu-Zeletin, R: Application of X.500 and X.700 Standards for Supporting Personal Communication in Distributed Computing Environments. - in: Proc. of the 5th IEEE Computer Society Workshop on Future Trends of Distributed Computing Systems, Cheju Island, Korea, August 1995

[11] Eckardt, T; Magedanz, T; Ulbricht, C; Popescu-Zeletin, R: Generic Personal Communications Support for Open Service Environments. IFIP World Conference on Mobile Communications, Canberra, Australia, September 1996

[12] Magedanz, T: TINA - Architectural Basis for Future Telecommunications Services. - in: Computer Communications Magazine, to appear in late 1996

[13] TINA-C Doc. No. TB_MDC.018_1.0_94: Overall Concepts and Principles of TINA. February 1995

[14] TINA-C Doc. No. TB_MDC.012_2.0_94: Service Architecture. March 1995

[15] TINA-C Doc. No. TB_HK.002_1.0_94: Service Component Specifications. March 1995

[16] Mowbray, Thomas J.; Zahavi, Ron: The essential CORBA, System Integration using Distributed Objects, Jon Wiley & Sons, Inc., 1995, ISBN 0-471-10611-9

[17] Eckardt, T; Magedanz, T; Schulz, M; Stapf, M: Personal Communications Support within the TINA Service Architecture - A new TINA-C Auxiliary Project. - in: Proc. of 6th TINA Conference, Heidelberg, Germany, September, 1996

[18] BERKOM Project "Intelligent Personal Communication Support System", Deliverable 0: Design criteria and preliminary description of functionality. - German National Research Center for Information Technology (GMD), Research Institute for Open Communications System (FOKUS), December 1995

[19] Deutsche Telekom Project "Personal Communications Support in TINA", Report No. 1. - German National Research Center for Information Technology (GMD), Research Institute for Open Communications System (FOKUS), June 1996

[20] Jayant, N.S.; Noll, P.: Digital Coding of Waveforms. - Prentice Hall: Englewood Cliffs, 1984

[21] Doermann, David; Yao, Shee: Generating Synthetic Data for Text Analysis Systems. - for: Proc. of the 4th Symposium on Document Analysis and Information Retrieval, SDAIR, Las Vegas, Apr. 24-26, 1995

[22] Jainschigg, John: Text-to-Speech. Sobering up the "drunken swede". - in: Teleconnect, May 1995, pp. 125-128

[23] Léwy, Nicolas; Hornstein, Thomas: - Text-to-Speech Technology. A Survey of German Speech Synthesis Systems. - UIBLAB Technical report 94.10.2, ftp://ftp.uiblab.ubs.ch/pub/paper/GermTTS.ps.Z

[24] Pfeifer, T.: Speech Synthesis in the Intelligent Personal Communication Support System (IPCSS). - in: Bateman, J. (ed.): Proceedings of the 2nd 'Speak!' Workshop on Speech Generation in Multimodal Information Systems and Practical Applications, 2. - 3. November 1995, Darmstadt, GMD-IPSI

[25] Rabiner, L.; Juang, B.-H.: Fundamentals of Speech Recognition. - Prentice-Hall: Englewood Cl., 1993

[26] Gales, M..J.F.; Young S.J.: Segmental Hidden Markov Models for Speech Recognition. - Proc Eurospeech '93, Berlin, pp. 1579-1582, Sept. 1993

[27] Hoefker; Hoehne; Jesorsky: Automatische Sprechererkennung mit Computern, BMFT Bonn, Forschungsbericht DV 80-002, 1980, cs: BMFT Bonn

[28] International Standard ISO/IEC 11172. Information Technology - Coding of moving pictures and accociated audio for digital storage media up to 1.5 MBit/s (MPEG-1). - Genf: ISO, 1993

[29] International Standard ISO/IEC 13818. Information Technology - Generic coding of moving pictures and accociated audio information (MPEG-2). - Genf: ISO, 1994

[30] CCITT Recommendat. H.261. Video codec for audiovisual services at p x 64 kBit/s. Genf: CCITT, 1990

[31] International Standard ISO/IEC 10918-1. Information Technology - Digital compression and coding of continuous-tone still images. Part 1: Requirements and guidelines (JPEG). - Genf: ISO, 1991

[32] Murray, James: Encyclopaedia of Graphics File Formats. - O'Reilly: New York, 1994

[33] Pfeifer, T.; Gadegast, F.; Magedanz, T.: Applying Quality-of-Service Parametrization for Medium-to-medium Conversion. - Proc. of the 8th IEEE Workshop on Local and Metropolitan Area Networks, Potsdam, Aug 25-28, 1996, publ.: Los Alamitos (USA), IEEE Computer Society Press

A Framework for the Deployment of New Services Using Hypermedia Distributed Systems

Álvaro Almeida

INESC, Praça da República, 93 R/C,
4007 Porto Codex, Portugal
tel: +351-2-2094200   fax: +351-2-2084172
e-mail: aalmeida@inescn.pt

Abstract. This paper presents some problems that arise in the deployment of new advanced telecommunication services and a new way to deal with them using hypermedia distributed systems. A special emphasis is placed on the propagation of the service in order to rapidly reach the subscriber/user.

1   Introduction

The enormous success of the World Wide Web, one of the most popular Internet services, brought to the public the wonderful world of the hypermedia and multimedia distributed systems. This success has put a great stress on service providers in order to satisfy the increasing user demand of new services.

The emerging and promising market of services puts under pressure not just service providers but also service developers and significant R&D efforts are being made to render more efficient the creation of new services. The European projects SCORE (RACE 2017 – Service Creation in an Object–oriented Reuse Environment) and BOOST (RACE 2076 – Broadband Object–Oriented Service Technology) made a large investment aiming at the rapid service provision and at the definition of a Service Creation Environment (SCE) [1], [2] to reduce the time spent on service creation. But there is still the need to reduce the time that elapses since the service become available until it is effectively used.

One could expect in the near future to find many new services running over different distributed platforms such as ANSA [3], DCE [4], CORBA [5], etc. This diversity of offers, consequence of a strong competition between distributed platform providers, can be an obstacle, in the eyes of the user, to the acceptance of new services. A solution is needed to make the customer understand that there is no reason to fear so many different service technologies. A possible solution is presented in this paper, based on hypermedia distributed systems.

This contribution is organised in 6 sections. In section 2, the timeline model, a model which shows the phases through which a service goes from its provision until its use, is presented as an overview to the remaining sections.

In section 3, some issues related with the services provision, particularly the obstacles to their rapid dissemination, are presented.

In section 4, a possible solution to the problems revealed in the previous section, is presented based on the hypermedia distributed systems. Also presented is the impact that the usage of hypermedia distributed systems to assist in the dissemination of new services will have on the service creation process.

In section 5, a case study that implements the proposed model is presented in detail.

Finally, in section 6, some conclusions are drawn about the issues discussed in this paper along with some leads for future work.

2   The service lifetime

In order to clearly identify the different phases of the service lifetime and to address the problems raised during each of them separately, the concept of Timeline Model was developed. This concept comes from an ETSI work [6] which summarises the views in a service lifetime of different timescales.

The model, illustrated in figure 1, shows that there is a hierarchy of timescales involved in the interactions between a user, a service provider and a network provider. The model presents three timescales: *i)* the service; *ii)* the contract; *iii)* the call.

i) The service

This timeline has a scale of decades and embraces the entire lifetime of a service. Starts with the *provision* of the service and includes the entire service creation process, from the initial studies about the future service until the service becomes available to be used. Consecutively, there is the service *operation* and, at the end, the *discontinuation* of the service when it is replaced or obsolete.

ii) The contract

This timeline describes the relations between the customer and the service provider. Starts with the *subscription* of the service by the customer, covers the *use* of the service and ends with the *cancellation* of the contract by either party. The timescale of this timeline spans days, months or even years.

iii) The call

This timeline examines the three stages of a connection–oriented call. Starts with the procedures of *access* to the service, which cover the user indicating that the use of the service is required, identifying himself to the provider and negotiating any parameters of the call. Consequently, there is the *information transfer* that covers the period of the effective usage of the service. Ends with the *disengagement* that concludes the process, generating and storing a charging record for the call and closing all the logs assigned to that user. This timeline can be measured in minutes or hours and the duration of each of its phases depends on the kind of service considered.

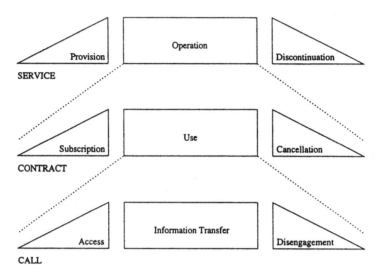

Fig. 1. The timeline model

3 Service provision issues

In a world-wide scenario, a citizen would be able to subscribe and use a service geographically placed anywhere. In order to use a service, he will need a service interface – client application – for each service subscribed. Even for similar services, involving the same media, a specific client application will be needed for each service either because they run over different distributed platforms or use different security mechanisms. Hence, the user will most probably need as many client applications as the number of subscribed services, or at least, as the number of service providers he deals with.

From a user point of view it is not acceptable to remember which client application to call for each subscribed service. Therefore, a mechanism should be provided to the user to allow him to use different services offered by different service providers in an easy way.

The exposed above shows three important issues that must be considered by the community involved in services provision: *i)* the supporting service concept; *ii)* the common user interface concept; *iii)* the multi-service framework concept.

3.1 The supporting service need

Strong efforts aimed to reduce the time spent on service creation have been made. However, one should bear in mind not just the time used to create the service but also the time needed to propagate the service in the market.

Consider the definition of $T_{p_{ss}}$ as the average propagation delay of the service, calculated as the mean time since the service becomes available for subscription (t_0) until the number of subscribers reaches the minimum number (U) that guarantees the economic feasibility of the service. This minimum number of subscribers has to be estimated during the initial phase of the service creation process, along with other studies made to evaluate the service feasibility. Those studies are necessary to decide whether the service will be created or not. In conjunction with the estimation of a minimum number of subscribers, an acceptable time interval to reach that value is also established. If the number of subscribers is less than the expected after this period, the service economic feasibility needs to be judged.

Figure 2 illustrates the propagation delay issue. t_i is the delay of propagation of the service to subscriber i. Thus, $T_{p_{av}}$ is represented as:

$$T_{p_{av}} = \frac{\sum_{i=1}^{U} t_i}{U}$$

Therefore, a mechanism which allows the service to rapidly reach the potential subscriber/user is needed. Considering that the potential subscriber/user could be, in many cases, anywhere, this mechanism should allow the subscription of the service remotely and 24 hours per day.

To use a subscribed service it may be necessary to download some pieces of software such as the service client application or the decrypt block. To allow remote subscription and tele–loading of software, a supporting service is therefore needed, which will assist in the propagation of new services.

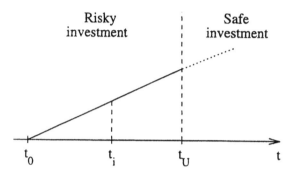

Fig. 2. Service subscription (in detail)

3.2 The common user interface need

In the near future the user will have the possibility to subscribe services provided by companies located anywhere around the world.

Since a hard competition between services to gain the same potential user should be expected, each service provider will try to captivate the user through an attractive interface with many functions and options. However, considering the expected number of services available for subscription, it may be difficult for the average user to remember which service client application he should call to access each subscribed service. The user will need a common application capable of relating the subscribed services with their own client application, as shown in figure 3.

The common application will activate the particular client application according to the user choice, which will then carry on its service connection tasks. This common application will act as a note–pad, pointing to the client application command from the user subscribed services list.

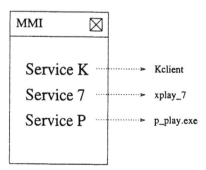

Fig. 3. A man-machine interface example

3.3 The multi–service framework need

The rapid evolving of services engineering and related technologies gives the perspective of a world-wide open market in which there will be a large number of service and service providers. A strong competition between service providers with several offers to the same kind of service should be expected.

Meanwhile, there is already a strong competition between distributed platforms providers. This competition will lead to a scenario where will be several services running over different distributed platforms.

A scenario with services using different data formats, different protocols, different security mechanisms running over different distributed platforms should be considered. On the user side, consequently, there will be a long list of applications to access each of the services. Those client applications will most probably use their own protocol to communicate and their own data format to encode the information.

Therefore, a framework is needed supporting multiple protocols and multiple data formats for each arising service on the user side. This framework has to be able to unify the available services from the user point of view. To the user there will be a single door to the services world.

4 Hypermedia distributed systems

Hypermedia distributed systems may offer an important help to new services, by providing a framework to their use and expansion. In fact, the hypermedia distributed systems can assist the propagation of new services, supporting the service subscription task and the tele–loading of the software necessary for their use. The hypermedia distributed systems give the chance to connect different services, using different protocols and different data encoding schemes. This feature makes the user believe that the services are inter–connected somehow, specially because it offers him a common interface to all of them.

4.1 Hypermedia distributed systems as a supporting service

In order to satisfy the high demand for new services, developers need to use methods and tools which help them to create services rapidly. The re–use of existing services components, together with object–oriented methods, such as the Object–oriented Modeling Technique (OMT) [7], the Specification and Description Language (SDL) [8] and C++ [9], proved to be a valid approach to the service creation process [10]. However, once created it is necessary to make the service reach its targeted user as fast as possible.

Hypermedia distributed systems can act as a supporting service assisting remote subscription of new services and tele–loading of the software needed by the user to access them. This software can be the client application to connect to the service or just a coded key, or even both. Acting as a supporting service, the hypermedia distributed systems help to reduce the service propagation delay. In an open market, the hypermedia distributed systems help the new service to reach its minimum number of subscribers rapidly, either because it allows the subscription 24 hours per day or because it increases the potential number of subscribers.

The framework model which joins a hypermedia distributed system (HDS) and a new service (NS) is shown in figure 4. During the subscription procedure the subscriber/user contacts the service provider using his HDS client application – Channel 1. The HDS server passes the request of the user to the NS server – Channel 2 – and continues the subscription task by sending the answer back to the user – Channel 1. This answer may include pieces of software. Once the subscription process is concluded, the HDS client application launches the NS client application – Channel 3 – which, in its turn, connects the NS server – Channel 4. During the service usage, channels 1, 2 and 3 are closed.

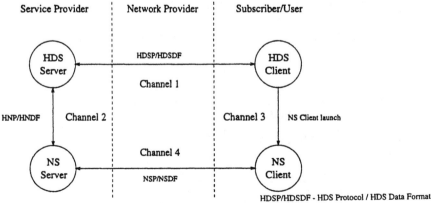

Fig. 4. The framework model

HDSP/HDSDF - HDS Protocol / HDS Data Format
NSP/NSDF - NS Protocol / NS Data Format
HNP/HNDF - HDS-NS Protocol / HDS-NS Data Format

4.2 Hypermedia distributed systems as a common user interface

The ability to deal with different multimedia objects and information servers, distributed around the world, allowing the user to navigate between them is the main feature of the hypermedia distributed systems. The hypermedia distributed systems link multimedia objects logically related and show them to the user, according to his requests, invoking the appropriate application to handle them.

Since we can define an object of service type and associate to it, on the user side, the service client application as the appropriate application to deal with it, we can create a list of services. Therefore, the user can have a list of his subscribed services which are linked to their own client application. This helps the user to catalogue the subscribed services and provides him with a friendly interface to all of them. The list will be updated with each new subscription.

Taking advantage of using a well known interface, the HDS client (figure 4), that provides the launching of the service client application, the usage of services becomes an easy task to any kind of user.

4.3 Hypermedia distributed systems as a multi–service framework

As described in section 4.2, the hypermedia distributed systems are capable of dealing with objects coded in different formats.

In figure 4 we can see the framework model, to connect a HDS to a new service. Extending the model in order to make the HDS deal with more than one service is straightforward. Therefore, a service provider can offer several services, all of them linked through the same HDS, which assist users in the subscription tasks.

Taking into account that a HDS server can be inter–connected with other HDS servers, since it is a distributed system, we have consequently all services providers, which use the same hypermedia distributed system technology, potentially connected. And, therefore, all the services assisted by the same hypermedia distributed system technology inter–connected in the eyes of the user.

Our door to the services world is then created, and the hypermedia distributed systems can in fact offer a multi–service framework, always ready to accept the newly arrived services.

4.4 Impact on the service creation process model

Consider, for instance, the model for the service creation process defined in the SCORE project. The SCORE process model [11] defines the following activities: Service Analysis, Service Specification, Service Design, Building Block Specification, Building Block Design, Building Block Coding, Service Implementation, Test and Deployment.

The Service Analysis activity aims to study the feasibility of the new service in a global view – technical, economic and legal. The results of all studies will determine whether the service will be created or not. During the Service Specification activity, the service is specified in an implementation independent way using formal methods. Based on the information gathered in the two previous activities, the service is design in the Service Design activity. The blocks identified at the design of the service are then specified, designed and created. During Service Implementation activity, the blocks are linked in order to create the service. Afterwards, the service is tested. Once the service succeed all tests, it is then deployed on the network infrastructure and becomes available.

The use of an hypermedia distributed system as the framework for the new service does not overload the creation process significantly. In fact, for some types of services it may even reduce the service creation effort.

The Service Analysis activity must include an additional study on the hypermedia distributed systems available. The study will focus on the evaluation of the number of potential users that it can bring to the new service and on the technical compatibility with the possible technologies to be used on the service creation.

The decision to create the service using an hypermedia distributed system does not force any other activity to change. It is necessary to create an additional block in the system, to connect the server block of the new service (NS) with the HDS server, and implement the HDS–NS protocol, as shown in figure 4. This fact affects the other activities just in the way that there is a new block, apart from the ones necessary to create the service.

Since this new block it is just used for service propagation purposes, it can be re–used from/to other services using the same hypermedia distributed system technology. A service provider will need only one HDS server running, if all the services he provides use the same hypermedia distributed system technology for propagation purposes.

5 A case study

A case study using hypermedia distributed systems to support the propagation of new services is being considered for cable TV networks. Taken advantage of the Internet Access service provided by the CATV operator to their users, TV channel subscription will be assisted by an hypermedia distributed system. For this particular case, World Wide Web was chosen to be the hypermedia distributed system.

The architecture of the system is illustrated in Figure 5. The subscriber will use his favourite WWW browser to accede the WWW server of his CATV provider. Once matched the user identification with the users database contents, the TV channels available for subscription are presented by the hypermedia system. The user will select the desired channel and will receive as an answer a decoding matrix with the decoding keys of all the TV channels he subscribes. This feature allows the use of the system to subscribe TV channels and to unsubscribe also. The user PC will pass the decoding matrix to the set-top box (STB) that decodes the TV channels. The TV decoding key for each channel is compound by an hardware part located at the STB and a software part sent by the hypermedia system during the subscription process. Therefore, the security of this system concerning decoding matrix transmission is guarantee. With this system, the user can subscribe and unsubscribe TV channels at any time, without the human interference on the CATV operator side.

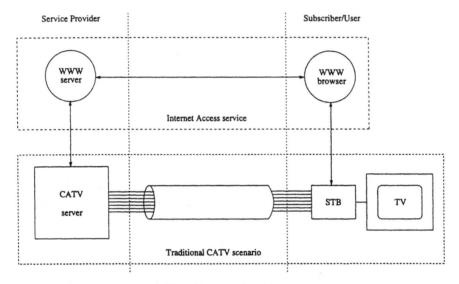

Fig. 5. The case study architecture

6   Conclusions

In this paper some problems were presented, with which the community involved in service provision have to deal, during the initial phase of the service operation – the service subscription phase.

The multiple offers that the user, in an open world-wide market of services, will have, involving several service providers adopting different technical solutions, can be an obstacle to the adherence to new services.

The hypermedia distributed systems can play an important role, assisting in the propagation of new services and giving the users a unified vision of the services market, without compromising the diversity of technical solutions involved in the services.

The case study presented is a partial implementation of the framework model. Traditionally, CATV operators use an hardware equipment called set-top box to decode the TV channels, which they supply to the customer when he signs the contract. This fact makes impossible to perform today the whole service subscription process using uniquely an electronic environment. Nevertheless, this case study proves that is possible to use hypermedia distributed systems to subscribe TV channels in a secure and ease way.

As continuation of the work presented here, we propose to analyse the use of hypermedia distributed systems to perform accounting and extra security tasks and look into the inter–connection of different hypermedia distributed systems.

References

1. R2017/SCO/WP2/P/028/b2, *SCORE - Methods & Tools*, Report on Methods and Tools for Service Creation (Second Version), Deliverable D204, December 1994.
2. R2076/MAR/MCS/DR/R/015, *Consolidation of the SCE Requirements and Architecture*, BOOST deliverable 15, July 1994.
3. Architecture Projects Management Ltd (ANSA Workprogramme), *System Programming in ANSAware 4.1*, ANSAware 4.1, doc. RM.101.02, February 1993.
4. John Shirley, *Guide to Writing DCE Applications*, O' Reilly and Associates, 1992.
5. Object Management Group, *The Common Object Request Broker: Architecture and Specification*, rev. 1.1, December 1991.
6. ETSI STC NA4 WP NA4.2, *The Timeline Model*, July 1991.

7. J. Rumbaugh, M. Blaha, W. Premerlani, F. Eddy and W. Lorensen, *Object-oriented Modeling and Design*, Prentice-Hall, 1991.

8. CCITTT, *SDL Specification and Description Language*, Recommendation Z.100, 1993.

9. Stanley B. Lippman, *C++ Primer*, Addison Wesley, 2 edition, 1991.

10. Álvaro Almeida, Paulo Proença and Manuel Ricardo, *MAD – a multimedia service over an ATM network*, Proceedings of the ATM Developments´ 95 International Conference, March 1995.

11. SCORE WP1, *The SCORE Service Creation Process Model*, Deliverable D104, Volume I – R2017/SCO/WP 1/DS/P/016/b2, RACE project 2017 (SCORE), December 1993.

ISABEL: A CSCW Application for the Distribution of Events

Juan Quemada, Tomas de Miguel, Arturo Azcorra, Santiago Pavon
Joaquin Salvachua, Manuel Petit, David Larrabeiti, Tomas Robles, Gabriel Huecas

Departamento de Ingenieria de Sistemas Telematicos (DIT)
Technical University of Madrid

Introduction

Many activities which in the past have required physical presence and direct interaction among participants can be performed in a distributed fashion with the help of advanced information technologies such as, CSCW [1,2] (Computer Supported Cooperative Work), interactive multimedia services and broadband communications. Technologies aiming at supporting the collaboration among individuals or groups are identified under the term groupware technologies. Asynchronous interactions which do not require physical presence of interacting persons have matured during the last years. Very successful examples of asynchronous groupware exist. LOTUS Notes [3] is considered probably the most successful commercial product in this area. The Internet and many of its application can be considered as groupware technologies to some extend.

Technology can support today also *synchronous* interaction where real time contact among individuals is required. We mean by synchronous interaction the exchange of verbal, visual, ... messages or information, like the exchanges of information carried out typicaly in meetings, conversations or other activities where several participants collaborate in physical presence.

Remote synchronous interaction is not new, the plain old telephone is a very good example of an old technology supporting a simple but very effective form of synchronous interaction. POTS is today by far the most demanded synchronous service. This service has evolved into N to N audioconference or videoconference facilities. Computers in general and the Internet have also had primitive types of character oriented synchronous interactive services for a long time, like TALK, IRC, ... Today low quality voice and video over the Internet is also common practice with applications like, CU-SeeMe, IVS, VAT, ...

Audio-visual broadcasting is also a highly demanded type of remote synchronous interaction which has been done since many years. Although broadcasting has really no interaction because the flow of information is unidirectional, it is nevertheless being addressed in the experiments performed for creating new synchronous services. One of the most popular services on the multicast backbone of the Internet, also known as the MBONE, is the conference broadcasting for which a Session Directory (SD) exists where the list of broadcast conferences is displayed in real time.

Sophisticated forms of remote synchronous interaction requiring good quality telepresence demand more bandwidth and more reliable communications to achieve a proper interaction. Therefore for setting up large sacle experiments like the RACE/ACTS Summer Schools [2,5,6,9] a complex collaboration among a large number of organizations has been needed. In addition, the availability of large broadband infrastructures is enabling the realization of large scale experiments using broadband communications which are providing a better understanding of the role which synchronous interaction will play in the communication services of the future.

Although intensive experimentation is needed, there are many application domains where remote collaboration seems to have the same usability as physical presence. This could avoid or reduce travel and/or movement of persons with substantial benefits for the overall productivity of an organization.

Introduction to ISABEL

ISABEL is a CSCW application whose very first version was developed in the European project ISABEL in 1993. The ISABEL project was leaded by TELEFONICA I+D. The other participants were CET, the research labs of Telecom Portugal, and the Dept. of Telematic Engineering of the Technical University of Madrid (DIT-UPM). The development continued in the RACE project IBER with the same participants. The acronym IBER was chosen due to the iberic nature of the project including only Spanish and Portuguese participants. The ACTS projects NICE, TECODIS and BONAPARTE have continued its use and tunning in a variety of application domains.

The ISABEL application was designed from the beginning for the interconnection of audiences. ISABEL grew associated with the realization of the RACE and ACTS summer schools on Advanced Broadband Communication, where several auditoriums had to be interconnected to create a unique large transnational distributed virtual auditorium, where the attendees achieve the sense of participating in a unique event independently of it´s location. The authors are not aware of other applications which have targeted at interconnecting audiences and claim ISABEL to be the first application to have addressed audience interconnection with integrated management of the distributed event. Related projects where audience interconnection are intended are : the Munin Project [8] which addresses lecture room interconnection and the ETSIT project addressing lecture room interconnection over satellite.

Interconnection of audiences imposes different HCI (Human Computer Interaction) needs than interconnection of desktops. Distributed events based on desktop interconnection are *heterogeneous* in nature in the sense that each participant has not the same view of the event. When using CSCW over desktops each participant in the distributed event shares with the rest only some of the windows he has on the screen and he adapts the view he has of the shared components to his particular needs.

We call distributed events *homogeneous* if all participants share the same view of the interaction. We believe that this is a fundamental requirement when distributing activities where the physical presence has naturally provided a homogeneous view of the event. Therefore when interconnecting meeting rooms, classrooms, .. a similar view of the interaction has to be provided in all endpoints. The homogeneity of views is the only way of achieving a similar perception of the activity by all participants and also of achieving the sense of participation in the same event.

ISABEL based Service Provision

ISABEL is a CSCW application which provides the basic technological framework for supporting remote collaboration in various areas of professional activity. ISABEL has been designed as a configurable CSCW environment which supports several interaction modes. The areas where ISABEL has been experimented and for which specific operation modes have been included are :

•Teleeducation/training : In education and training the use of tele-lecture-rooms could facilitate access to education in remote regions or in locations where the expertise is not available.

•Telework : In the working environment the need to facilitate the communication and collaboration between teams is considered necessary to improve productivity.

•Telemeeting : The meeting is a central management element. The creation of telemeeting rooms which avoid travelling can substantially change their role. Now meetings are scheduled according to travelling constraints, whereas telemeetings can be scheduled according to project needs shortening lead times and increasing productivity.

The experience gained from experimentation shows that in all the above mentioned activities several types of interactions appear during an event. Therefore, the introduction of different interaction modes in a CSCW application provides a much more natural understanding and feeling of the course of the event to the remote participants. Lets take teleeducation as an example of the large variety of interaction which can appear during the course of an event. We can differentiate different kinds of events and also different types of interactions in those events. For example :

•Telelecture : A tele-lecture-room covering a large geographical area, where the students can participate in the lecture from the distance. The interactions occurring during a lecture will be basically two: 1) a presentation part usually based on viewgraphs or a blackboard ; 2) a questions/answer part.

•Teleseminar : A seminar or brainstorming session where ideas are presented and discussed. A brainstorm is more unpredictable than a lecture, but it will also consist of sequences of interactions like: 1) verbal discussions among several participants ; 2) discussions based on sketches on a blackboard ; 3) discussions based on last minute documents ; 4) more formal presentations based on viewgraphs ; ....

•Teleconference : Conferences, symposiums, workshops, ... are ussualy massive events where participants follow the state of the art of their particular fields of knowledge. As the rest of events, the conference consists of a sequence of interactions like : 1) a presentation part usually based on viewgraphs, slides or photographs; 2) a questions/answer part ; 3) Panel discussions ; 4) demos ; ...

•....

The approach used when designing ISABEL has been to define an architecture, where different interaction patterns can be created and experimented. Service creation with ISABEL will therefore imply the identification of the interactions to be supported in order to create a new service mode. Support to a new service is introduced in ISABEL by creating a new management/operation mode which enables an effective control of the required interactions.

Elements of ISABEL

As in most CSCW application the collaboration process is supported in ISABEL by having WSs or PCs where a distributed multimedia application allows users to share elements or media by providing a coherent view of them on all the computer's screens. We call *sites* to the physical places where the WSs or PCs are located and where participants join to a given distributed event. A distributed event consists therefore in a sequence of intercations performed through the ISABEL application among several sites. A service provision platform which includes, workstations, audiovisual equipement and a proper communication subnetwork is used to run ISABEL.

Three conceptual parts can be distinguished in ISABEL :

•Telepresence. The achievement of sense of presence of the remote participants by means of audio and video transmission. ISABEL supports the most general telepresence paradigm : videoconference from N to N participants. Up to 16 sites have been connected with simultaneous video transmission from all to all.

•Shared Workspace. A shared media space which enables users to achieve a common view and understanding of the objects or ideas subject of the collaboration. Each mode, teleconference, teleeducation, telemeeting, telework, ... has a different shared workspace in ISABEL.

•Interaction control. The means by which an ordered collaboration is achieved in a conference with remote participants. In ISABEL interaction control acquires the functionality of a management environment which has full control of all the elements of the application if needed. Each different application scenario, teleconference, telemeeting, telework, ... is supported in ISABEL with a different management scheme which supports the interaction roles of the group (hierarchical, brainstorm, flat, etc.).

In homogeneous events the WSs or PCs all sites will have always the same layout with the same information presented in order to provide to all participants the same view of the event. In heterogeneous events this restriction is not necessary. ISABEL supports both kinds of events.

The creation of each particular service platform will require a special mixture of those 3 elements adapted to the particular HCI characteristics of each case. For example a teleconference service which has to interconnect large auditoriums will have completely different requirements than a telework service designed to connect individual workers on their workstations. Therefore, the interaction between ISABEL components can only be understood on a service by service basis and will be explained in a later section. We will provide now only a general overview of the main components and characteristics of ISABEL.

Telepresence

Telepresence is achieved by audio and video transmission among sites such that participants can see and hear each other despite of the distance. In order to achieve a natural interaction some scenarios require N to N video transmission. Each interaction mode in each particular service has usually different HCI requirements and therefore ISABEL has a modifiable telepresence component which can be tailored to the specific needs of each interaction type.

In each interaction there are active and passive participants. Usually the active members of the configurations are the audio and video sources which participate in the videoconference among sites. For example, in the lecture mode only the video of the speaker is distributed, whereas in a panel discussion the videos of all the panellists are distributed.

The audio and video components of ISABEL have the following characteristics :

•Audio: This component can provide audio channels among all the partners involved in the conference. It supports different bitrates and qualities ranging from 8 bit PCM to 16 bit CD quality. Audio codecs which compress the audio bit rate are also supported in ISABEL. In particular, GSM and G721 codecs are supported. This wide range of audio codings have been included for experimentation purposes and to allow adaptation to different communication channels.

•Video facilities: This component provides video connections among all the users involved in the conference. ISABEL can present the video images of all participants in the conference in all sites. Up to 16 simultaneous video images have been performed. The video component uses MJPEG compression. The reason for this choice is just the availability of high performance video boards supporting this algorithm for the SUN WSs used.

Shared Workspace

The Shared Workspace is a shared media space which enables users to achieve a common view and understanding of the objects or ideas subject of the collaboration. The shared media space is very dependent on the application domain, tele-education, tele-work, tele-meeting, ... The shared environment must be able to present to all users a common view of the problems or elements subject of the collaboration. For tele-education conventional lecture support material like slides, pointers or blackboards must be shared, whereas in tele-work arbitrary applications running, pointers or documents must be shared.

Examples of media components existing in the shared workspace of ISABEL are :

• Viewgraph presentation: This component allows one participant to control a viewgraph based presentation from one of the sites to the rest of the slides.

• Pointer/Pencil: This component allows one or more participants to use pointers and pencils to point or draw in other components of the shared media space.

• Editor: This component allows users to produce documents in collaboration.

• Whiteboard: This component provides a distributed graphical editor which enables the joint creation of graphic designs, diagrams, flow charts, etc.

• Display sharing: This component can present part of the workstation display at one site to the rest of participating sites.

• Window Fax: This component allows to capture any part of the screen and send it over to the other participants of a collaborative session.

The exact configuration and accesibility of the different components of the shared media space is definable for each type of interaction by means of the interaction control component of each service mode.

Interaction Control

Interaction control, also called floor management, is the key component which permits a natural and easy use of CSCW. Interaction control is the means by which an ordered collaboration is achieved in a conference with remote participants. The central idea behind the ISABEL interaction control component is that the management and operation of distributed events must follow a similar approach as the one used in the original non distributed event.

For example, the ISABEL based management and control of the virtual auditorium created in a teleconference by interconnecting several auditoria has been conceived as an extention of existing auditorium control rooms. Auditoria have control rooms from which the audiovisuals are controlled according to a script which is prepared in advance. An operator controls, open microphones, audio levels, video cameras, projection, lights, ... according to the script during the course of the event. Likewise, a virtual auditorium is controlled in ISABEL by a common script which defines all the interaction modes to be used. An operator in the central event control room will change the interaction modes through the ISABEL teleconference control panel available in the event control workstation.

In telemeetings the management scheme has particular constraints. In addition, a large variety of meeting types exist. The telemeeting service of ISABEL provides support to a conventional chaired working meeting. In it, a chairperson runs the meeting and decides the activities to take place in the meeting on the basis of a given agenda. The management of a telemeeting service must be controlled here by the chairperson. The chairperson must decide about the interactions to take place on a more on the fly way than in a teleconference. No central control room is needed nor convenient now.

ISABEL was developed to support the RACE/ACTS Distributed Summer Schools on Advanced Broadband Communications [5,7,10]. Those summer schools targeted from the beginning at auditorium interconnection. Therefore complex event management issues appeared from the beginning. This led to an application architecture which has a programmable management/control part which has been used to experiment with management variants and modes. The basic characteristic of ISABEL is that all telepresence and shared workspace components arc dynamically programable. The interaction control part is built in TCL-TK and allows an easy prototyping of new control functions.

With the experience gained with the distributed summer schools several management modes have been introduced which enable effective management of distributed events with a large number of sites. The fourth summer school ABC´96 can be used as an example of its capabilities. It took place in July 1996 and is the largest distributed event supported with ISABEL. In ABC´96 up to 18 sites were interconnected in 14 countries (Austria, Belgium, France, Germany, Greece, Iceland, Italy, the Netherlands, Norway, Portugal, Spain, Sweden, Switzerland) on two continents. ABC´96 lasted 4 days and lecturers were hosted in 6 different auditoriums (Belgium, Canada, Germany, Italy, Portugal and Spain). The event control center of ABC'96 was located in the auditorium control room of the Technical University of Madrid site. Without a centralized control scheme such a large event would have been imposible.

Teleconference Service

Conferences, workshops, seminars are meetings which have required traditionally physical presence. A teleconference service must enable the interconnection of physically separate auditoriums in order to create a unique virtual auditorium where participants can interact with any other participant in any auditorium. A teleconference is clearly an homogeneous event where all the participants must have the same view of the distributed conference. In this mode the usual types of interactions among participants must be supported, such as, lectures, talks, demos, panel discussions, question sessions, ...

From the cost point of view a teleconference service can be seen as a service platform available to a large population of potential users. The operations cost of the platform will be shared therefore by the users. Although the cost of the platform can be high, the cost will be amortized in many events. The cost of distributing the conference will be shared also by many participants. Therefore the individual share of the cost of distributing a conference can be affordable even with today's prices.

In order to understand a distributed conference it is convenient to describe it from two different views : the participant´s view and the control view. Both are complementary.

The Participant´s View

A distributed conference can be modelled as a set of participating sites, each having an associated role and functionality. A site is an access point to the virtual auditorium for participants and consists usually of a physical auditorium which is connected to the rest of the auditoriums by the projection of the screen of a multimedia WS where the ISABEL application is running. Several types of sites can be differentiated.

First of all we have *Interactive Sites* (referred as *IS*) with a maximum interaction functionality. From the organizational point of view some ISs can be differentiated due to the fact of holding lecturers or having a registered attendance. Such sites have been called *Main Sites* (refered as *MS*) in the ACTS/RACE Summer Schools. MSs have ussualy the same functionality as ISs but must provide a more reliable service. On the other extreme we have *Watch Points* (referred as *WP*) which are receive only sites. There may exist also uncomplete interactive sites which provide to participants only a limited set of interaction facilities. For example a site which can not support the realization of presentations by the lecturers but which can pose questions. The next figure ilustrates the structure and components of a distributed conference.

The core of the conference is formed by a set of ISs, including some MSs, interconnected ussually by an ATM multicast subnetwork. This core constitutes the interactive part of the conference. Each interactive site will have an auditorium which can interact with the other auditoria according to the mode set by the event control center. Additionaly ISABEL can be used to set up Watch Points. A watch point has exactly the same view of the interaction as the interactive sites, but it can not interact with the rest. A watch point needs only a unidirectional connection from the core network to the site. ISABEL can de used also to set up gateways to the MBONE.

The set of ISs of a given event form the core of a distributed event, whereas the WPs are only able to follow what is happening in the ISs. During the course of a distributed event the interactive sites follow a sequence of interaction patterns which are derived from the script and content of the event. In each interaction among the

N interactive sites of the event, there will exist A active sites and P passive sites. The active sites of an interaction deliver information content defined by the role they are playing, such as a speech, slides, video images, computer demos,... to all the participants in the event. Of course the roles and attributes of the ISs change when the interaction mode changes. For example : During a lecture, usually only the site where the lecturer is present is active and has the lecturing role which enables the speaker to control the shared components needed for his presentation, like viewgraphs, pointer, pencil, display sharing,... When the talk finishes and questions are raised, the sites which pose questions get active and are able to send audio and video. In a panel discussion all the sites which host a panellist are active sites and the rest passive.

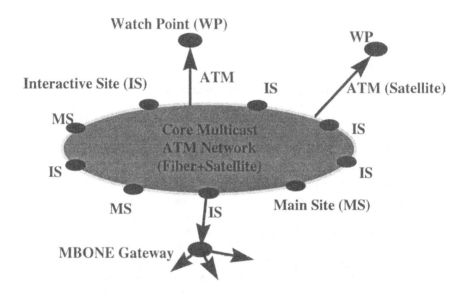

The Control View

The change of roles and attributes in the core of the distributed event is performed by the operation team in the *Event Control Center*. As in any non distributed conference or TV production studio, local control rooms which selects the right microphones, cameras, lightening,... must also exist in all the auditoriums participating in the distributed event.

The next figure illustrates the control model used in ISABEL for a distributed conference. Each site has two ISABEL WSs. The first WS is used to project the shared media space and the telepresence components in a large screen in the auditorium. From this screen, the lecturer controls his presentation. The second WS, also called the control WS, is used to control and manage de distributed event. This acts as the conference creator and all sites must connect to it when joining the event. The rest of the control WSs have only a marginal role in the event, related to the adjustment of the local audioinput to the ISABEL audio component. Control workstations are ussually located in the control rooms of the auditoriums. Of course, local control of audio, video, microphone, video cameras,.. must also be performed in each auditorium complementing the overall interaction mode control performed in the event control center.

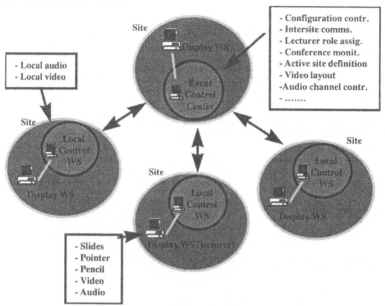

The control WS in the event control center can select the interaction mode established between the sites with the control panel of the interaction control component which is depicted in the next figure.

This control panel controls the configuration of all the WSs participating in the distributed event. Pressing a button in this control panel creates a chain of signals to all the workstations participating in the event which will configure in each WS the appropiate layout and state. This panel includes several types of buttons. For example the middle column is used to select interaction modes such as:

One site in large (1 large video)
Two sites interacting (2 videos)
Three sites interacting (3 videos)
Show all sites (all vieos)
Presentation/lecture mode (with three video sizes and several locations on the screen)
Question-answer mode

The lower part of the includes some audio and video related controls. The right side column is used for selecting the presentation which is going to be given. The left side column is used for selecting the sites of the interaction. Finally the row on the bottom is used for adjusting the bandwith used by ISABEL.

The ISABEL operation manual describes in detail the precise meaning of each button.

A properly designed conference should include a script which precisely defines the sequence of changes. Each change is performed ussualy by pressing one (or more) button.

There exist more control and monitoring panels in the event control center. Those are described in the operation manual.

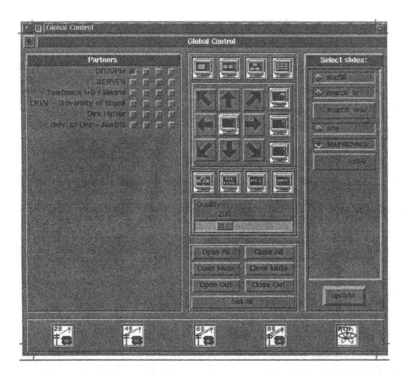

Telemeeting Service

Ideally, a telemeeting service should interconnect geographically disperse meeting rooms where the attendees can interact among them as if they were in a single room. There are many kinds of meetings which are potential candidates for being transformed into a distributed meeting and not all can be covered with a single service. The ISABEL telemeeting service supports to a conventional chaired working meeting. Such a meeting has two conflicting requirements. It must support, on one hand, the chairperson running the meeting and provide him the instruments to control the interaction modes when needed. On the other hand, it must facilitate an unplanned and easy change of interaction modes as decided by participants, in order to encourage active involvement of the participants in the meeting. Therefore the control of the telemeeting shall be in the meeting room and not in a separate control room as in the teleconference.

The ISABEL telemeeting falls into the category of homogeneous event where all the participants must have the same view of the event. Its main target is interconnecting meeting rooms, but it can be also used to connect directly from a workstation to the distributed event.

From the cost point of view a telemeeting service can be seen as a service platform available to a large population of potential users. A regular telemeeting service should be based on a set of telemeeting rooms available to projects which should be booked and used by projects. The operations cost of the platform will be shared therefore by the users. Although the cost of the platform can be high, the cost will be amortized in many telemeetings. The cost of the meeting will be also shared by the participants. Therefore the individual share of the cost of distributing a meeting can be affordable with today's prices.

The Participant's View

A distributed meeting can be modelled as a set of participating sites, each having an associated role and functionality. Each site is an access point to the virtual meeting room for participants and consists usually of a meeting room where a WS connects through a multimedia application to the rest of the sites. If the number of attendees to a meeting room is large a screen projector can be used. Usually all sites will be interactive (*IS*) with full interaction capabilities. Although in some particular meetings *WPs* may play a role, this is rare and we will deal only with telemeetings formed of interactive sites.

Some of the sites may be individuals sitting in their WSs and attending to the meeting as shown in the picture.

During the course of a distributed meeting the sites will pass through a sequence of activities, each having a particular interaction pattern. The chairperson shall have the possibility of taking the last decision about the activities performed, but he is also a participant and must therefore be able to participate as anybody else. Typical interactions occurring during a meeting are : 1) A speech from one site ; 2) A viewgraph based presentation ; 3) A discussion among some or all sites ; 4) the distribution and discussion of a document ; 5) To show a demo ; 6) To discuss a paper ; 7) ...

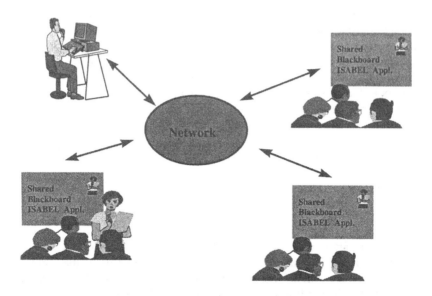

The Control View

The change of roles and attributes of the sites must be performed in a natural way. On the other hand the chairperson has to have the last word about the decisions taken. The ISABEL telemeeting mode tries to harmonize both requirements by giving to the sites the capability of requesting resources and roles directly.

The control panel of the ISABEL telemeeting mode appears in the same WS where the telepresence and shared media components are presented. Therefore only one WS per site is needed to set up a telemeeting platform. The next figure shows the control panel used to interact in a telemeeting , which has been made as simple as possible to facilitate its use by any participant with a short training.

The interaction rules are as follows : To facilitate the interaction the audio channels are always open such that any participant can always communicate verbally with the rest of the sites. The controls allows the change of the video layout and the creation of elements of the media space, like pointers/pencils, presentations, shared display, notepad, blackboard,...

The following functions are supported in the control panel:

1) Set big video : the video image of a site is set in large.

2) Set debate mode : The video image off all sites participating in the telemeeting are presented in the largest size possible. This interaction mode should be used for discussions where only voice is used.

3) Set intermediate mode : The video image off all sites participating in the telemeeting are presented in a medium size such that a large part of the screen is still free for the shared media space. This interaction mode should be used when video image is important, but other components of the shared work space should be used.

4) Set work mode : The video image off all sites is presented in a small size at the righ hand side of the screen. This interaction mode should be used when other components of the shared work space should be used, such as shared blackboard or shared display.

5) No video mode : all the videos are removed from the screen. This interaction mode has been included for the case that participants would like to remove all video information from the screen.

6) Set presentation mode : All the elements of a presentation are assigned to the site requesting it. This includes : presentation video in the upper right corner, presentation selection panel, slides, and pointer/pencil.

7) Shared display : A window is created by the site requesting it and the window is displayed in all the remaining sites.

8) Pointer/pencil : A pointer which can point and draw in any window or part of the screen is created. The pointer can be used by all participants.

9) Shared Blackborad : A shared blackborad where all the sites can draw and interact is created.

10) Shared Notepad : a text editor where all the sites can write.

Network Issues

The availability of event specific broadband multicast subnetworks is the main limiting element for the set up of distributed events. Events with two sites have some usefulness, but the real impact of distributing events comes from minimizing movements of persons. Therefore service platforms must be scalable and support an increasing number of participating sites. Multicast is the only effective way of sharing the bandwidth. The cost of establishing a full mesh network grows quadratically and this is unacceptable for more than two sites.

Experiments performed [16] show that the traffic aggregation in a CSCW application is much lower than the sum of the individual peak traffics of the sources, due to the strong correlation existing among shared media sources in the most frequent interaction modes. Therefore the bandwidth needed for the realization of a distributed event is not too large. For example, ABC'96 was a distributed event with up 18 interactive sites. An ATM multicast network was created for it with 6Mbit/sec connections. The average traffic was under 3 Mbit/sec and the quality was good. This was achieved with an MJPEG compression for video signals. As video is the most bandwidth demanding signal, a higher compression rate algorithm could have reduced the bandwidth needs.

Different approaches can be taken for creating a broadband multicast network. Let's describe briefly three of them which have been used for distributing events.

To perform multicast at the application level with the help of application routers and by using only unicast communications at the network level. This approach has the big advantage of needing only standard point to point connections at the network level. WSs are cheap and can perform copies of application packets with a reasonable efficiency when the number of copies is not very large. It is a very cost effective choice when the performance obtained allows its use. In the 1994 summer school this approach was used to interconnect 5 sites with ISABEL [7].

To use an IP multicast service like the MBONE. This approach works fine, except that the tests performed with some of the existing routers have shown a quick saturation of routers with a small number of copies (4 or 5).

To perform multicast at the ATM level. This approach is technologically very effective as most ATM processing operations. Multicast nodes making over 10 copies have been successfully setup for ABC´96.

The lack of efficient and flexible multicast servers which facilitate the creation of tailored multicast zones is one of the main hindrances for setting up distributed events. In the experiments performed for setting up the ACTS/RACE distributed summer schools, the deployment of broadband multicast subnetworks connecting the sites participating in the event took a very large share of the total effort.

Network architectures supporting distributed events not only have to adapt to the heterogeneous quality of end-point equipment. Furthermore, when the available bandwidth is not constant throughout the network, the quality of service at the application level or even the functionalities have to flexibly and smoothly adapt to the QoS provided by the network. In practice this implies the existence of QoS-adaptors connected to the backbone multicast network supporting the event at high quality, and taking the job of reducing the quality of service or the number of components transported in the multimedia stream towards the sites being serviced fitting each link's bandwidth. This enables easily scenarios with users connected through narrow-band access networks.

Conclusions and Further Work

ISABEL is a multimedia CSCW application which was created associated to the realization of distributed events. The experience and understanding gained during the four summer schools shows a significant progress but there seems to be room for further additional expansion of this cost effective approach to training in future summer schools or other educational events. The 1993 Summer School was the seminal event with only two auditoriums interconnected. In 1994 five auditoriums were connected. For ABC´95 ten sites were connected. ABC´96 has been the largest event organized with ISABEL with 18 interactive sites connected trough terrestrial ATM links and satellite. Additionally 5 ISABEL watch points have been receiving the summer school with the same quality as in the interactive sites. Growth in terms of size seems unpractical if interactivity among the sites is demanded. Of course, with broadcasting an unlimited growth seems feasible. The advancement in the following years is expected in functionality and reliability.

The experience gained shows that management of distributed multi-conferences is a central aspect in distributed event organiztion. Proper models identifying the roles of each actor at each site have to be developed and clear procedures have to be defined. Future developments which further develop the ISABEL interaction control part are planned for the future.

Very significant improvements in the quality of the interaction among attendes are notizable from the first versions of ISABEL. Nevertheless technology is advancing very rapidly and incorporation of many new elements like better components, more effective video codecs, audio codecs or echo cancellers,... is considered convenient to achieve a more interaction.

References

[1]   P.Wilson: "Computer Supported Cooperative Work". *Computer Networks and ISDN Systems*, pp 91-95, volume 23 1991, North Holland.

[2]   Ralf Steinmetz: "Multimedia: Advanced Teleservices and High-Speed Communication Architectures", Ed., Springer-Verlag - *Lecture Notes in Computer Science*, Volume 868, September 1994.

[3]   L. Lindop, T. Relph-Knight, K. Taylor, A. Eager, G. Einon, K. Joyce, A. Stevens: "Groupware : Let's work Together". *PC Magazine*, August 1995.

[4]   S. Pavón, T.P. de Miguel, M. Petit, J. Salvachua, J. Quemada, L. Rodriguez, P.L. Chas, C. Acuna, V. Lagarto, J. Bastos: "Integracion de Componentes en la Aplicacion de Trabajo Cooperativo ISABEL", Jornadas Telecom 94, Noviembre, 1994.

[5]   T.P. de Miguel, S. Pavón, J. Salvachua, J. Quemada, P.L. Chas, J. Fernandez-Amigo, C. Acuna, L. Rodríguez, V. Lagarto, J. Bastos: "ISABEL - Experimental Distributed Cooperative Work Application over Broadband Networks", pp 353--362, Springer-Verlag - *Lecture Notes in Computer Science*, Volume 868, September 1994.

[6]   Arturo Azcorra et al.: "Multicast IP support for distribuited conferencing over ATM". INTEROP 95. Las Vegas (USA).

[7]   A. Azcorra, T. Miguel, J. Quemada, S. Pavon, Department of Telematics from UPM, P. Chas, C. Acuña, P. Aranda, Telefonica I + D, V. Lagarto, J. Bastos, J. Domingues: "Distance Learning: Networks and Applications for RACE Summer School '94", Centro de Etudos de Telecomunicacoes, *The ATM Forum Newsletter* September, 1994 - Volume 2 Issue 3.

[8]   K. Age Bringsund and G. Pederson: "The Munin Project", Research Report, University of Oslo, 1993.

[9]   J. Quemada, T. Miguel, A. Azcorra, S. Pavon, J. Salvachua, M. Petit, J. I. Moreno, P. L. Chas, C. Acuna, L. Rodriguez, V. Lagarto, J. Bastos, J. Fontes, J. Domingues: "Distribution of ABC'95 over the European ATM Pilot Network with the ISABEL Application", Broadband Islands Conference, Dublin September 1995.

[10]   J. Quemada, T. Miguel, A. Azcorra, S. Pavon, J. Salvachua, M. Petit, J. I.
        Moreno, L. Chas, C. Acuna, L. Rodriguez, V. Lagarto, J. Bastos, J. Fontes, J.
        Domingues: ABC'95: A Tele-education Case Study", High Performance
        Networking for Teleteaching - IDC'95, Madeira November 1995.

[11]   J. Quemada, T. Miguel, A. Azcorra, S. Pavon, J. Salvachua, M. Petit, J. I.
        Moreno, P. L. Chas, C. Acuna, L. Rodriguez, V. Lagarto, J. Bastos, J. Fontes,
        J. Domingues."Tele-education Experiences with the ISABEL Application",
        High Performance Networking for Tele-teaching - IDC'95, Madeira
        November 1995.

[12]   Carolina Cruz-Neira, Daniel J. Sandin, Thomas A. DeFanti: "Surround-Screen
        Projection Based Virtual Reality : The Design and Implementation of the
        CAVE".Annual Conference on Computer Graphics, ACM-0-89791-601-
        8/93/008/0135, 1993.

[13]   Michael E. Luckacs, David G. Boyer: "A Universal Broadband Multipoint
        Teleconferencing Service for the 21st Century", *IEEE Communications
        Magazine*, November 1995, Vol. 33, No 11, pp36-43.

[14]   Jose Encarnacao, Martin Gobel, Lawrence Rosenblum: "European Activities
        in Virtual Reality", *IEEE Computer Graphics and Applications*, January
        1994, pp 66-74.

[15]   Jeffrey HSU, Tony Lockwood: "Collaborative Computing".*Byte*, March 1993.

[16]   José I. Moreno: "Propuesta de Arquitectura de Red de Banda Ancha para
        Servicios de Telecomunicación de Trabajo Cooperativo". PhD Thesis. Dpto.
        Ingeniería de Sistemas Telemáticos. Universidad Politécnica de Madrid. Junio
        1996.

The Bookshop Project:

An Austrian Interactive Multimedia Application Case Study

Helmut Leopold and Richard Hirn
Alcatel Austria AG
Scheydgasse 41
A-1211 Vienna
AUSTRIA
E-mail: [hleopold][hirn]@aut.alcatel.at

Abstract

As field trials all over Europe show, Interactive Multimedia Services (IMMS) are usually considered only for the market of private households with services like 'Video on Demand' and 'Teleshopping' via TV and Set Top Box in the living room at home.

Pilot examples in Austria demonstrate that there are other ways to use the same technology to realize interactive multimedia services for private and public companies and institutions. In this respect, the customer premises equipment need not to be necessarily a TV set but might be a PC representing a 'Point of Information' or 'Point of Sale' terminal located in offices, in stores or even in schools. Calculations based on experiences with an Austrian retailer who realised an interactive application predict that by such an approach an economic break even can be reached within 2 years. In addition, a broader distribution in the market is reached much sooner than by following the approach for the households mass market.

1. Introduction

The present day telecommunications market is faced with a challenging new environment - particularly in the provisioning of Interactive Multimedia Services (IMMS) which are based on leading-edge technologies that will be attractive to users. The market is changing at such a speed that traditional business procedures are not able to respond fast enough. In addition, provision and deployment of these new services involve high levels of complexity, encompassing various traditionally separated industrial sectors and areas of expertise.

Thus the information technology industry, the telecommunication industry and the entertainment industry are beginning to cooperate in new ways, both with each other and with end-users, in order to be able to respond successfully to the evolving market requirements and to explore new business opportunities. European projects in the framework of the European RTD[1] programs and national activities like the

---

[1] RTD ... Research, Technological development and Demonstration.

„bookshop" project as described in this paper will significantly contribute to the understanding of the necessary cooperations, the technological realisation and the economic and social aspects of new IMMS services within an European framework.

The involvement of Austrian industry in such projects integrates Austria into the European RTD framework and activities on advanced communication services based on broadband communication. Alcatel Austria is aiming to establish key pilot projects in this area to ensure that Austria plays a role in the establishment of the „European Information Infrastructure (EII)" [Tay96, ETS95a].

This paper summarizes the background, the basic motivation and the technical realization of the „bookshop project" in Austria. The bookshop project is a trial implementation of an Interactive Multimedia Service (IMMS) within a dedicated application scenario of a retail company. Section 2 summarizes the technological status that allows us to realize sophisticated multimedia systems for implementing IMMS. Section 3 provides the main reasons why trial activities especially for IMMS are an essential prerequisite for further development. Section 4 gives an overview of the main goals, the detailed application scenario and the technical approach of the bookshop project. Section 5 summarizes the major outcome of the trial and section 6, finally, describes the intended future work and the possibilities to utilize the system platform of the bookshop project in further application areas.

2. Interactive Multimedia Services as a new Business Area

The liberalisation in the telecommunication market and thus the advent of competition has a tremendous impact on business in this area. For example in the telephone market, new international network operators and CATV network operators will start to offer telephone services in competition to the classical national network operator. This will have an impact on both the market share and the tariff structure. A way to maintain or increase revenue is to additionally offer new services to the customers. The provisioning of new multimedia services [Leo94a] is made possible by the availability of several new technologies:

1. Optical signal transmission and SDH technology in the core network is the basis for the powerful ATM technology which allows the dynamic establishment of high-bandwidth ATM connections.

2. ADSL for twisted-pair telephone lines and HFC technology for coaxial CATV networks allows to utilize existing access network infrastructure in order to offer broadband services additionally.

3. Low cost digital mass storage renders economical the deployment of digital video servers.

4. Powerful video compression (MPEG-1, MPEG-2) effectively reduces storage expense and transmission bandwidth demand.

Network operators operating in a market that is deregulated already (e.g. British Telecom, Deutsche Telecom) have started to investigate this new market of IMMS. Meanwhile, pilot activities have been started all over the world [Grif96, Canc96, Liv96].

3. Importance of IMMS Trial Activities for Europe

3.1. General Impacts

It is essential to contribute to the establishment of the European communication community as follows:

1. Interactive Multimedia Services (IMMS) represent a new RTD area that has major technological, economic and social impacts and are seen as one of the most promising business areas especially for the public network framework. It is essential that European industry engage in this new evolving business area.

2. IMMS trial activities combine the interests of the research community, network providers, service providers and equipment providers on a national and international scale, contributing essentially to the establishment of an European Information Infrastructure.

3. Trial activities are a pre-requisite for the stimulation of awareness and for the identification of the potential economic benefits of new communication services based on leading-edge technologies in general and on interactive multimedia services in particular by the practical demonstration approach. This is the basis for realizing Interactive Multimedia Services (IMMS) in new business fields like industry, trade, consulting/remote learning, services like bank, insurance, etc.

4. The use of leading edge technology (broadband communication) to provide advanced telecommunication services and applications to potential end-users will stimulate the broadband market in general.

5. Experiences of IMMS trial activites will have benefits for technology and content providers, for network operators, for the end-user community and, most important, for many SMEs which may find new activities in the IT business (application designers, advertising and promotion industry, etc.).

3.2. Economic and Social Impacts

Interactive Multimedia Services (IMMS) are the basis for the provisioning of new applications in order to bring the right information to the right places at the right time (examples being 'Information On Demand', 'tele-working', 'tele-teaching', 'News on Demand', etc.). Appropriate solutions in this area will have a tremendous social and economic impact, as follows.

New remote working environments (at or near the home) will have social, economic and environmental impacts - changing working patterns, reducing regular travel thus reducing travel costs and environmental damage, and possibly introducing new business opportunities. Training and professional updating within companies will be much more efficient, as staff can be supported with up-to-date and just-in-time information. New methods will be developed for the remote education of people at all levels, including school children and college pupils, and also for people at their homes, where a mass market may be expected to develop.

However, it is a conclusion of the Austrian pilot activities guided by Alcatel Austria, that IMMS applications in the business area will be the first to be implemented and to justify the implementation investment. But it has still to be noted, that other services that meet the people's unlimited appetite for information of all sorts will surely develop and grow: this will include 'News on Demand', public and Government information services, library access etc. and finally entertainment services like 'Video On Demand (VOD)'.

Finally, the provisioning of new IMMS applications in Europe -- in an environment of liberalisation in the telecommunication market -- will contribute to the competitiveness of European industry.

4. Summary of the Trial Activity in Austria

4.1. General

The trial phase of the bookshop project which took place in Vienna from August 1995 to January 1996 was aimed at the demonstration and evaluation of technical feasibility and user acceptance of new interactive broadband telecommunications services by implementing one dedicated application in a typical networking scenario. It was also the intention of the trial phase to investigate the applicability of the standards and architectures developed in the ITU-T, ATM Forum, ETSI, ISO and DAVIC.

A key issue for the deployment of IMMS in the business environment is the clear understanding and the interworking of the various partners in a liberalized market environment (i.e. service provider, network provider, content provider, etc.). In addition, these parties need to be supported by appropriate Customer Care and Billing (CCB) systems. Generic facilities have to be developed to provide a framework which allows the different providers (network, service, information) to apply their individual tariff structures.

Thus the major objectives of the trial activity were to establish a starting-point for investigating the following issues:

1. to identify and define economic and cost-effective systems for the provisioning of interactive digital multimedia services for the business and the private sector;

2. to describe a generic system platform and to identify and implement generic system components which provide unified access to new IMMS services;

3. to define a framework and architecture for the interworking between the various partners for the provisioning of interactive MM services;

4. to develop schemes and mechanisms for the collection of charging information which are generic and cover the needs of all parties involved (i.e. network operator, service provider, information provider, customer) in a liberalized market;

5. to test the use of the service platform by implementing selected applications;

6. to investigate the user acceptance and to analyze and identify the impacts and requirements of new IMMS services on public network infrastructures;

7. to cover the end-user needs and acceptance of new interactive multimedia services.

The telecommunication developments in the multimedia area will be mainly guided by user and application requirements [Leo94b], and by economical feasibility aspects. Thus the emphasis of the trial phase was on stimulation and testing of new market fields. For the initial trial project, the most important factors in designing the trial network, realizing a dedicated application scenario which is described below, were the

- establishment of a cost-effective network scenario (based on the available public network infrastructure), and

- the use of available technologies on the market.

4.2. Application Scenario: „Bookshop Project"

One of the largest retail corporations in Austria has stimulated the IMMS trial activity. The high street shops of this retail corporation include supermarkets, drug-stores, and paper & bookshops. The corporation structure consists of one headquarter in Vienna and many outlets distributed across Austria. The whole logistics is managed from the headquarter. The basic motivation for the use of new communication technologies in general and of IMMS applications in particular in such an organisation can be summarized as follows:

1. providing information on-demand for the employees at the outlets (via dedicated terminals for the employees);

2. product presentation, advertisement, etc. to the customers at the outlets via public Point of Sales (POS) Terminals (see below);

3. stock optimization;

4. virtual supermarket: combining the IMMS with an appropriate ordering system implements a virtual supermarket where goods are offered which are not in stock which bypasses and compliments the outlets;

It is expected, that the use of IMMS in such an application scenario will lead to:

- employee work load reduction;

- up-to-date information for employees;

- up-to-date information for customers;

- increase in sales volume without significant manpower and resource increase in the outlets;

- creation of new market segments by virtual supermarkets.

The overall system architecture and the technical approach for the realization of such an application scenario is described in the following sections.

User Interface: Point of Sales Terminals

Home entertainment services like VOD will usually involve the TV-set as the end-user's terminal, while business services (teleworking, etc.) will see the PC as an appropriate terminal. If information is to be provided to the public, Point Of Sales

(POS) terminals and Point of Interest (POI) terminals, respectively, will be applied. The user interfaces of such terminals must be particularly self-explaining and simple.

Bookshop Trial Service description

The service was offered during the usual shopping hours with the following characteristics: POI terminals are used, comprising a TV screen, sound and a small keyboard. The menu-based application guides the customer through the product palette offered. To each article, some information can be retrieved by the user. Depending on the article, this is either a video-clip (in case of a CD), a trailer (in case of a video tape), or a short presentation. The user can control the play-back by usual VCR commands: play, pause, stop, fast forward, and rewind.

4.3. Technical Approach

General System Architecture

The basic network architecture for the bookshop project is based on the specifications of international standardization (ETSI, ITU-T, ATM Forum, DAVIC, etc.) [ETS95b].

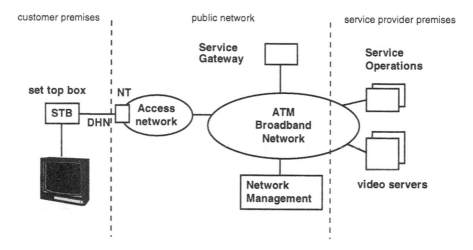

Fig. 1: IMMS Network Architecture

The network can be divided into three main sections (see Fig. 1) [Del94,Cha94,Fur95]:

1. The customer section (customer premises), comprising the private section of the network (DHN - digital home network - in the future), the set top box (STB) and the terminal equipment (TV set or PC).

2. The network provider section, including the core network (ATM broadband network), the access network and the network terminations (NT) (it has to be noted that the NT can be located at the customer premises). The Service Gateway (which was not implemented in the trial) allows the customer to choose among the services offered by different service providers.

3. The service provider network section, including the video servers, service operations server and a local ATM network.

- *Video Servers*: The video servers contain the video programs (usually MPEG encoded multimedia information). The video servers are owned by service providers. Multiple service providers may operate within the same network.

- *Service Operations*: The service operations servers finally enable the service provider to run the service he offers. These servers maintain service, traffic and customer information in databases and cover tasks such as:

  – control of the application, depending on the database contents (e.g. presentation of products depending on the stock, for the example of a warehouse application);
  – provision of text information extracted from the database, that is presented within the application (e.g. price information);
  – customer care and billing (CCB);
  – management of multiple video servers, which are belonging to the same service provider, distributed across the network, which may be in different distance to the customer and may contain different parts of the multimedia information requested;

It is important to note, that the various functional blocks, described above, are usually owned by different organisations.

Possible Access Technologies

The expense for upgrading a telecommunications network in order to support broadband services is mainly concentrated on the access network section. Depending on the physical media used for the access network, different technologies will be employed. According to [ETS95b], the main options are:

1. twisted-pair copper lines using ADSL technology;

2. Coaxial cable using DVB standard;

3. Hybrid Fiber Coax (HFC) networks with different levels of fiber deployment (typically the fiber is terminated in a node that serves a few hundreds of subscribers);

4. Fibre to the building or fibre to the curb (FTTB/FTTC) networks using copper twisted pair or coax to cover the last section to the terminal equipment;

5. Fibre to the home (FTTH) with a point-to-point or a point-to-multipoint configuration (passive optical network -PON);

6. 155 Mbit/s B-ISDN fiber access links.

The access network for the Austrian IMMS trial conformed to (1): twisted pair copper lines. For this access network architecture, in conjunction with an operable ATM core network, following network architecture is state-of-the-art (Fig. 2):

All connections within the service provider premises and the core network are based on ATM. ATM may be terminated in the Customers Premises Equipment (CPE) or within the access adapter.

The technology of the access network is ADSL, allowing most efficient utilization of the twisted pair copper subscriber lines. The asymmetrical bandwidth offered by ADSL matches the bandwidth demand scenario imposed by interactive multimedia services, where the required bandwidth is clearly higher in downstream direction. A further advantage of ADSL is the possibility to simultaneously use the same subscriber line for POTS and broadband traffic (which is not the case if using HDSL for example).

As shown in Fig. 2, an ADSL system consists of an access adapter and a number of Network Terminations (ADSL-NT). The access adapter, which is located in the local exchange (LEX) of the public telephone network, contains one or more network terminations (NT) towards the ATM network and a number of line terminations towards the subscriber line (ADSL-LT). Each line termination serves several, e.g. four, subscriber lines. As mentioned above, the ADSL-NT may be located either at the edge of the public network or at the customer premises. In Fig. 2, the ADSL-NT is within the customer premises, directly interfacing to the STB.

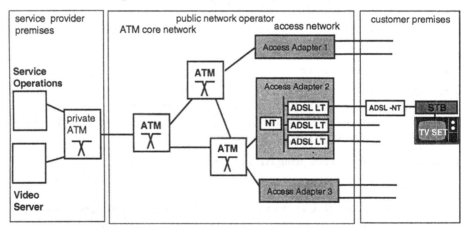

Fig. 2: IMMS Network Architecture - ADSL Access Network

It is important to note, that the technical approach followed for the Austrian trial activity is in line with the technical approach used by the VOD trial of British Telecom [Liv96].

4.4. Actual Trial Configuration

The network architecture as shown in Fig. 1 and Fig. 2 was the basis for the design of the trial configuration. However, the configuration actually implemented was not fully in line with this architecture due to a couple of external pre-conditions, mainly due to economic guidelines. In more detail, following constraints had to be considered:

- For economic reasons, the infrastructure of the public network operator's (Austrian PTT) network should be changed as little as possible. The trial should be realized using existing services, such as 2 Mb/s and 9.6 kb/s leased lines. It was out of scope to introduce ADSL equipment into the public network.

- The public network is structured in hierarchical levels, with the CPEs being connected to the core network via local exchange offices (LEX). The six terminals which were operated in the course of the trial were geographically dispersed over an area several kilometers in diameter. The locations were served by different LEX offices. Neither of these offices had an ATM access to the Austrian ATM backbone network.

- Even if there had been the possibility to use ADSL, this would have been economically disadvantageous, since for topological reasons each single ADSL-NT would have required a complete ADSL Access Adapter at its dedicated LEX.

Following the constraints mentioned above, the commercial available 2 Mbit/s E1 leased lines were used to transport the video information from the video server to the CPE (MPEG-2 downstream) and the 9,6 kbit/s leased line service was used to transport control information from the CPE to the service operations (up stream).

The system used only the upstream direction of the 9.6 kb/s link (downstream control data was multiplexed into the E1 MPEG-2 stream). Therefore, both the E1 and the 9.6 kb/s leased line, which are inherently bi-directional, were utilized in one direction only. Of course it would have been possible to bridge the 9.6 kb/s upstream connection over the 2 MB/s leased line, but at this time no proper equipment was available.

In future integrated systems, both downstream and upstream channel will be transported over ATM anyhow, which makes the transport of both information flows within a common physical link in the access network an easy task.

The actual configuration therefore looks as shown in Fig. 3. As can be seen, the configuration does not comply with the desired aim to provide ATM bearer service up to the customer premises. In fact, the ATM network was confined within the premises of Alcatel Austria (service provider premises).

The access adapter (not an ADSL, but a conventional ATM service multiplexer) is located at the service provider premises. The access adapter terminates the ATM protocol and provides a number of combined E1/V24 interfaces towards the public network. Those interfaces each consist of a bi-directional RS422 9.6 kb/s (also incorrectly referred to as 'V24') asynchronous serial port and a unidirectional E1 G.703 (2 Mb/s) port for downstream traffic.

From the access adapter, leased lines across the public network form transparent connections to the terminals at the customer premises. Besides the six terminals within the city, one local terminal was located next to the service multiplexer.

Fig. 3 shows that HDSL modems are employed in the access network, forming a part of the infrastructure for the 2 Mb/s leased line bearer service. Within the core network, the 2 Mb/s data was transported over a combination of SDH and PDH multiplexers. Similarly, the 9.6 kb/s traffic is transported over conventional modems, also using twisted-pair copper lines.

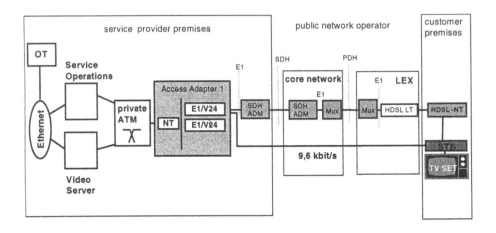

Fig. 3: IMMS Network Architecture - actual trial configuration

The multimedia information, the STBs' operating system and the application software are stored on one central video server at the premises of Alcatel Austria. The server has a storage capacity of 10 GB. The audiovisual information is compressed according to the MPEG-1 standard, with an overall bit rate slightly below 2 MB/s. Both audiovisual data and software are further encapsulated in an MPEG-2 transport stream (MPEG-2 TS) at 2.048 Mb/s. The data is stored onto the server in this format.

During operation, the server will stream out these files over the 2 Mb/s leased lines. Yet empty packets within the streamed data may be replaced with control information, thus allowing the server to control the STB over the 2 Mb/s stream.

At the service provider premises, a dedicated Ethernet segment is used to connect the video server with the service operations server, transporting control information between these devices. Ethernet is also used to connect the Operation Terminal (OT; a SUN workstation) of the video server.

4.5. ATM Information Flows

All ATM connections are semipermanent connections, involving an ATM cross-connect configured by management operations. Looking at the ATM level, following ATM information streams are used as shown in Fig. 4:

1. *Downstream*: ATM connection between video server and access adapter for the transport of the video and audio stream (movie), boot and application data and control information.
2. *Upstream*: ATM connections between service operations server and access adapter for the transfer of upstream control information.

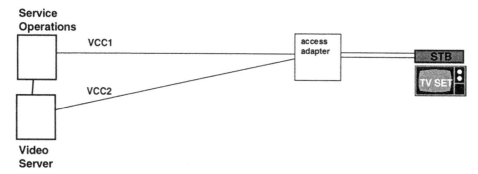

Fig. 4: ATM Connections

The protocol stacks used for the transport of all types of downstream information are sketched in Fig. 5 and Fig. 6.

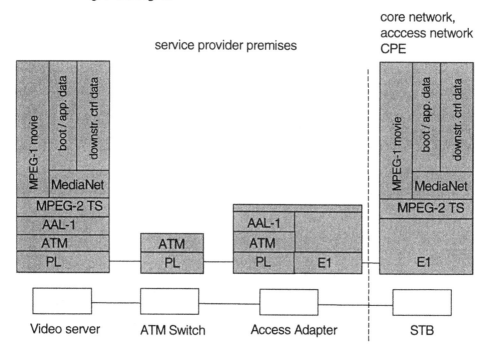

Fig. 5 Information Flow - downstream direction

As shown in Fig. 5, all downstream information is multiplexed into an MPEG-2 transport stream (MPEG-2 TS). By the use of the AAL-1 service, this information is transferred via the ATM network to the access adapter, and then without a further transport protocol to the STB.

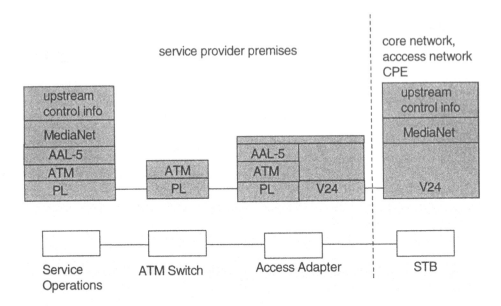

Fig. 6: Control information flow - upstream direction

As shown in Fig. 6, the upstream control information is transferred using the 9.6 kb/s line (here referred to as V24).

Both the STB software (boot and application data) and the control data (both directions) are exchanged over the Oracle MediaNet protocol.

5. Major Results of the Activities

One of the goals of the trial was to establish a network infrastructure which allows the demonstration of a dedicated IMMS application. Thus, the main focus of the trial was on the application itself. Valuable information could be gained on following issues:

1. The technical feasibility of a 'true-interactive multimedia system' comprising several POI (point of interest) / POS (Point of Sales) - terminals has been experimentally verified.

2. The concept of sending a 2 Mb/s data stream over cascaded ATM and E1 networks was tested. Although this should not be much of a problem, it revealed some difficulties.

3. The operation of such a system under real-life conditions was tested during a period of several weeks (the application was running and accessible to the customers during the usual shopping hours).

4. By continuous monitoring of customer interaction, information could be gathered allowing conclusions about customer acceptance. Furthermore, the customers' interest in the POS terminals and their understanding of the application could be observed.

5. Ergonomic and safety aspects of the POS terminals could be studied in an environment comprising children and physically handicapped customers, as well as unskilled operation personnel.

Important outcomes were:

1. At the current technological state-of-the-art, interactive multimedia systems comprising a telecom network are technically feasible. However, the expense of getting such a system running should not be underestimated. Due to its high complexity, the system will be extremely sensitive to changes in topology or physical transmission media.

   It turned out that the protocol used for data transfer needs some tune-up in terms of error handling and throughput performance. Any inconvenience during transmission caused the application to crash. Operation could only be resumed by hard-resetting the STB. Since the staff at the outlets were reluctant to do this several times a day, we finally had to construct a remote-controlled reset circuit to reboot the STBs from the service provider's site.

2. The employed concept of combined AAL-1 and E1 transport (in contrast to an ATM end-to-end solution) revealed several standardization inconsistencies in the field of jitter and wander specifications. As a consequence, severe transmission problems had to be dealt with, especially in conjunction with dynamic start and stop of video streams. Two basic problems could be identified:

   First, an ATM network may introduce a cell delay jitter of up to 1 ms. At the E1 interface of the service multiplexer, the jitter has to be bound to some microseconds, depending on the jitter frequency (according to ITU-T G.823). It turned out that the adaptive clock recovery mechanism implemented in our ATM service was not capable of reducing the jitter by this extent with the given buffer size. Especially at the start of AAL-1 transmission, it is by principle not possible for the service multiplexer to instantly reconstruct the correct clock rate from the rate of the incoming cells. The reason is that even tiny cell delay jitter will massively change the observed cell rate, if the observation period is small. Increasing this period means increasing the buffer size. Unfortunately, buffering expense and delay would become uncomfortably high, if the E1 jitter requirements shall be complied to even under worst-case assumptions (1ms jitter within the ATM network).

   Second, an E1 leased line is usually supposed to transport a data stream of constant clock frequency, stemming from one and the same source all the time. In an interactive TV network, it may happen that the E1 stream is generated by different sources alternatingly, or that the source is switched off occasionally and the service multiplexer fills in an alarm indication signal. In these cases, the clock frequency may change at the transitions within a range of +/-50ppm, according to ITU-T G.703. It turned out that it would take the E1 network several seconds to resynchronize after such disruptions, which caused massive data loss and bit errors. This caused massive problems in the STB startup phase, where the video server sent out the STB boot image immediately after taking up generation of the AAL-1 stream. To get the system running, we therefore had to make the video server insert several seconds of idle data before starting transmission of important data.

Fortunately, such problems will not occur (at least not within the transmission network) as soon as end-to-end ATM transport is provided.

3. During operation, system stability turned out to be an important factor for POI terminals. Unlike home entertainment systems, no customer interaction can be expected to reset a terminal once the application hangs.

4. The customer does not care whether the POS terminal represents the endpoint of a sophisticated network or it is a stand-alone device. It is only the user interface that counts, unless the customer is aware that he really needs the actuality of the contents provided.

5. To succeed, the IMMS applications have to be highly user-friendly and should be very simple. The use of a keyboard as an input device showed severe ergonomical drawbacks. Most users were not able to identify the correct key to press for the desired function. The need to implement a POI/POS terminal with a touchscreen rather than a keyboard to ease operation became obvious.

6. A matter of high importance in conjunction with user friendliness is appropriate response time of the system. The actual response time of three and more seconds did not go well with the behavior of the customers, who randomly pressed one key after the other to get a reaction of the system. The optimum response time to cope with this behavior must be well below half a second. If the time necessary to e.g. interrupt a movie and start another one cannot be decreased to fall below this limit, an additional progress indicator (like Window's hourglass) must be introduced to show the user that the response to his request is in progress.

7. IMMS applications are not only for the non-business market (e.g. house holds). It was clearly identified, that IMMS applications for the business market are of the same or even greater importance than in the non-business market. Calculations based on this particular retailer business application scenario in Austria predict that with such an approach an economic break even can be reached within 2 years. Thus, a broader distribution in the market may be reached much sooner than by following the approach for the households mass market.

6. Future Work

The functions presently implemented (i.e. the detailed trial network) were considered to be the minimum set that support IMMS applications. Other additional functions are of course necessary to make the system ready for future applications; i.e. to offer the flexibility, QoS, and other system support for future demanding applications, thus following the recommendation of international standardization (most importantly taking into account the DAVIC results) and implementing a full end-to-end ATM solution. It is intended to build on the experiences of this trial project and to consider the results of European RTD projects in this field (e.g. the ACTS project OPARISoD) for the realization of a powerful IMMS system. The most important application area in the future is seen in business applications like:

- public administrative authorities
- education: tele-learning in environments like schools, universities and companies;
- tourism - hotels: information and ordering systems in hotels, tourist areas, etc;
- banking: new possibilities to offer bank services to the customers;
- health: use of such a system for the exchange of medical information;

Thus new trials will be planned taking into account the experiences of the bookshop project and considering new achievements like the availability of an Austria wide public ATM network.

7. References

[Can96] E. Cancer, G. Freschi, „The Italian Interactive Multimedia Trial", Proceedings of the 35th European Telecommunications Congress, Vienna, Austria, August 27th- September 1st 19996.

[Cha94] Y. Chang, D. Coggins, D. Pitt, D. Skellern, M. Thapar, C. Venkatraman, „An Open-System Approach to Video on Demand", IEEE Communications Magazine, May 1994.

[Del94] D. Deloddere, W. Verbiest, H. Verhille, „Interactive Video on Demand", IEEE Communications Magazine, May 1994.

[ETS95a] ETSI, „Report of the Sixth Strategic Review Committee on European Information Infrastructure". European Telecommunication Standards Institute, Sophia Antipolis, June 1995.

[ETS95b] ETSI/NA5, DTR/NA-52109 „Video on Demand Network Aspects", Technical Report, October 1995.

[Fur95] B. Furth, D. Kalra, F.L. Kitson, A.A. Rodriguez, W.E. Wall, „Design Issues for Interactive Television Systems", COMPUTER, May 1995.

[Grif96] M. Griffith, F. Guirao, L. Van Noorden, „Network Evolution for residential Broadband Interactive Services - From RACE to ACTS", European Conference on Multimedia Applications, Services and Techniques (ECMAST '96), Proceedings Part I, May 1996.

[Leo94a] H. Leopold, K. Frimpong-Ansah, N. Singer, „From Broadband Network Services to a Distributed Multimedia Support-Environment", In Proc. of the International COST237 Workshop on Multimedia Transport and Teleservices, D. Hutchison, A. Danthine, H. Leopold, G. Coulson (eds), Vienna, November 1994 (Springer Verlag LNCS 882, ISBN 3-540-58759-4).

[Leo94b] H. Leopold, A. Campbell, D. Hutchison, N. Singer „Distributed Multimedia Communication System Requirements", ESPRIT Research Reports, Project 5341, OSI95, Volume 1, A. Danthine (ed), The OSI95 Transport Service with Multimedia Support, 1994 (Springer Verlag, ISBN 3-540-58316-5).

[Liv96] A. Livingstone, „BT Interactive TV", Proceedings of the 35th European Telecommunications Congress, Vienna, Austria, Aug. 27th-Sept. 1st, 1996.

[Tay96] M. Taylor, "Creating a European Information Infrastructure", TELECOMMUNICATIONS, January 1996.

Issues in the Design of a New Network Protocol*

Mikael Degermark[1] and Stephen Pink[1,2]

{micke, steve}@cdt.luth.se

[1] CDT/Department of Computer Science, Luleå University, S-971 87 Luleå, Sweden
[2] Swedish Institute of Computer Science, PO Box 1263, S-164 28 Kista, Sweden

Abstract. We describe some of the issues in the design of a new packet switched network protocol. Adaptation to various network technologies along the dimensions of speed, error model, robustness, etc., is a goal for this new protocol. We look at the adaption in size of the packet header to the speed and robustness of the underlying network to allow efficient communication on low-speed wireless networks, for example. We also explore issues in resource reservation and multicast for real-time multimedia, the notion of a network "flow", a hybrid of datagrams and virtual circuits, and suggest common solutions for both mobile and multicast routing. The authors are engaged in the design of a network protocol, NP++, whose goal is flexibility over a wide dynamic range of speeds and varying kinds of hardware switching elements.

1 Introduction

The last quarter century has produced mainly two sorts of packet-switched networks: virtual circuit and datagram. Virtual circuit networks emulate circuit switching by setting up a permanent path through a network for packets to travel during the time of a connection. Datagram networks allow packets to be routed along any available path such that no two packets have to follow the same path in the network. There are clearly advantages and disadvantages to both of these schemes. In an effort to provide the most flexible protocol for next-generation networks, we are developing a new network protocol that we call NP++. This paper will describe some of the issues in the design of this new protocol.

A new network protocol must be flexible enough to operate over a large variety of communication technologies. It must be designed for flexibility in the face of networks with widely different speeds, latencies, error characteristics, packet formats, abilities to provide quality of service, etc. A new network protocol should build on the core properties of both datagram and virtual circuit networks, as well as on the the properties of their hybrids.

* This work was supported by a grant from the Centre for Distance Spanning Technology (CDT), Luleå, Sweden, and a grant from Ericsson Radio Systems AB

2 High *and* Low Speed Networks

The new generation of computer networks can be characterized as having a wide dynamic range of properties. Speed, latency, robustness in the face of errors, and other properties, vary much more among today's networks than last generation's. New high speed fiber optic networks are being developed at the same time as relatively low speed radio channels. Such high speed networks exhibit very low error rates especially in comparison with lossy wireless links.

Today's network protocols are being designed mainly with high speed links in mind. This is generally a good guiding principle since even low speed radio channels will be getting faster. But protocol designers should also realize that the *dynamic range* across the set of new networks will also be increasing. Thus, flexibility to adapt to both high and low speed networks, high as well as low delay paths, and high and low error rates, should be the design principle for a next generation network protocol.

A new network protocol should adapt to such a wide dynamic range of computer networks. Borrowing from a number of traditions in the history of packet switching protocols, the new protocol should take advantage of the most successful features of both datagram virtual circuit protocols with the goal of minimizing their disadvantages. For example, where connections and state in the network make sense for communication, the new protocol should maintain a virtual circuit path between sender and receivers; in other situations, it should use datagrams as the main mode of packet transmission. Similarly, on networks with low bandwidth–delay products, a new protocol should rely on end-to-end mechanisms at the transport layer to control congestion. On high bandwidth–delay product networks, hop-by-hop congestion control should be provided. This is because the minimum granularity for the end-to-end control of congestion in a network is one round-trip time, and the amount and rate of change of data in a high bandwidth–delay product network may be too large to control during this time.

A key idea in NP++ is that a packet header (or trailer) is seen as a set of independent fields which can all be sent with each piece of data, or can be sent separately from other header fields more or less frequently. Thus, by removing large addresses from packets and assigning small, temporary connection identifiers, header compression can be achieved on a low speed link with soft state that is refreshed by the arrival of address field chunks sent less frequently than the data [10, 12]. On a high speed link the small connection identifier may speed up forwarding as no routing lookup is necessary.

3 Network Hardware Architectures

Another dimension in which NP++ is designed to be flexible is in its adaptability to various network hardware architectures. Network switches and routers are being developed in rapid succession to provide high performance switching, quality of service such as bounded delay, etc. New hardware switch designs often require changes to packet formats, forcing redesign of the protocol when new switch architectures are deployed. Advances in VLSI design methods, for example, have enabled the production of some low-cost, high speed switching fabrics for small fixed length packets (cells). Thus, it has been argued that today's ATM cell switches offer the best price/performance for high speed packet switching. On the horizon however are new designs and prototypes for IP routers [19] that will provide cost-effective routing of variable-length packets at multi-gigabits per second, where large packets increase the total throughput.

The point is that new network hardware architectures are being developed at an increasing rate and we should not have to specify, negotiate, deploy or even have to *choose* new network protocols each time next-generation network hardware is introduced. So, it is important for a flexible network protocol to adapt to new hardware in the network. We are designing a network protocol that can be either *switched* as fixed size cells or *routed* as variable length packets depending on the hardware of choice in the network. For this to be the case, the new protocol must be able to fragment and reassemble itself in the network when necessary [15].

4 Datagrams and Virtual Circuits

This section describes the two dominant ways of designing protocols for packet switched networks: *datagrams* as in the IP protocol used in the Internet, and *virtual circuits* as in ATM.

When a source is sending packets into the network, it needs to specify the source and destination of the packets. With datagrams this is done on a per-packet basis and with virtual circuits it is done once per connection. The source address is needed for error reporting, some forms of multicast forwarding, and perhaps cost allocation. The destination address is needed for locating and/or specifying (as in the case of a multicast address) the intended receiver(s).

4.1 Datagrams

Datagram protocols place full addresses in every packet, which means that routers do not have to keep any *inter-packet state* for routing purposes[3]. This has obvious advantages in terms of robustness and simplicity since packet forwarding state need not be maintained in the network. For example, if the path a stream of packets is following through the network fails, packets will automatically detour around the damaged part as soon as the dynamic routing protocols find a viable path.

If the address space is large, however, the lookup required to find the forwarding information can be costly. Today the size of a backbone Internet routing

---

[3] Some datagram networks use source routing, where each packet contains the complete list of routers the packet should pass through on its way through the network. We do not consider such networks here.

table is around 35,000 routing entries.[4]

Moreover, routing information is not the only kind of information that may be needed to forward a packet. If some packets need special treatment, for example, to get acceptable delay through the network for real-time voice and video, routers may have to maintain and access forwarding state for those packets. For each packet, a router must determine if the packet needs special treatment and, if it does, the router must access the forwarding parameters for that kind of packet.

Current work in the IETF on Controlled Load Service [28] is attempting to reduce the amount of forwarding state needed by lumping packets together into *classes* of packets that are treated in the same way. Such aggregation does not reduce the lookup time, however, since the size of the lookup domain is not decreased. Classification of packets is typically done on the basis of addresses, transport protocol, and port numbers. Aggregation of the form discussed in [28] reduces the amount of forwarding state by sharing it, but does not help to simplify the procedure of deciding what class a packet belongs to. In a general classification scheme, where it is possible to specify special treatment by arbitrary combinations of addresses and port numbers, classification can be an expensive operation.

4.2   Virtual Circuits

Network protocols built on the virtual circuit model do not put full addresses in each packet. Instead, a connection phase occurs before data is sent in which addresses are given to the network and routing and forwarding state is established in the network switches. The switch normally uses a small integer per-hop to access the forwarding state, and this index is made consistent at switches on both sides of the hop. This index is sometimes called a *virtual circuit identifier, (VCI)*, and is placed in each data packet belonging to the connection. In addition, there has to be an end-to-end virtual circuit identifier that uniquely identifies the circuit across all of its hops, as in ST-2 [26]. These two identifiers, along with the per-switch identifiers that must be present in order to negotiate the VCI between any two hops, mean that virtual circuit networks need at least three levels of naming. An advantage of virtual circuits, on the other hand, is that it reduces forwarding state lookups to simple table lookups based on the VCI. This advantage works especially well when there are many packets sent per connection.

One problem for virtual circuits is that one must establish a connection before transmitting any data, even in the case where only a single packet needs to be sent. This adds complexity and delay. To alleviate this problem for virtual circuits, various caching schemes to reuse existing connections and reduce setup delay have been developed. Another problem is that the forwarding state is typically *hard* and must be torn down by explicit control messages, which adds

---

[4] Statistics of the routing table at the Mae East NAP are available from http://compute.merit.edu/stats/mae-east/routing-table . Its size (prefix count) on Aug 20th 1996 was 34617 entries, and on Oct 10th it was 38515 entries.

further complexity. *Soft-state* [5], where timers and refresh messages keep state in place avoiding tear-down messages, has been developed for networks where hard-state is a problem.

In large networks, routing tables are usually maintained by dynamic routing protocols. Changing network conditions can cause such protocols to modify the routing table. With virtual circuits such changes do not affect the quality of service since the path through the network is pinned down. With datagrams, however, changes in the routing table causes packets to follow different paths through the network, possibly resulting in changes in service quality. Routing changes are not uncommon in the Internet. A recent study [20] shows that the routing instability in the Internet has increased slightly between December 1994 and December 1995. When the network is actually damaged, on the other hand, virtual circuit networks require special mechanisms if connections through the damaged part are to be recreated on paths circumventing the damage. If such special mechanisms are not used, the failure of a single router along the path of a connection will destroy the connection. With datagrams, all paths between sender and receiver must fail before communication fails.

4.3 Datagrams or Virtual Circuits?

With both datagrams and virtual circuits a router needs to access a routing table to be able to figure out where to forward packets. In the datagram case, this routing table may have to be accessed once per packet. For virtual circuits, the routing table is accessed when a connection is established. If all IP addresses were given out systematically such that all addresses in a certain part of the Internet had identical prefixes [16, 24], the number of entries in routing tables would be smaller than today, probably in the hundreds instead of in the tens of thousands.

The cost of a routing lookup would thus be significantly reduced and there would be one less advantage of virtual circuits over datagrams. The routing lookup would cease to be the bottleneck of router processing. Changing IP addresses in the current Internet to achieve such small routing tables is not an option. The required changes to millions of computers and routers around the globe are administratively and politically impossible. However, new datagram network protocols, for example IPv6, are designed to allow simple renumbering of addresses and can achieve such fast routing lookups. In such networks the virtual circuit model offers no advantages to traditional elastic data traffic and it is clear that such traffic is best served by the datagram model.

To support real-time traffic, future routers are likely to maintain many entries of forwarding state related to quality of service. Thus, it is packet classification that is the processing bottleneck instead of the routing lookup. The feature of virtual circuits to associate forwarding state with a small identifier in each packet can be important to allow fast lookup of such forwarding state. Traditional bulk data traffic, however, does not need quality of service handling and can be forwarded with less processing as only the fast routing lookup is needed.

In a flexible network protocol all packets should be able to carry complete addresses. They can also carry an identifier to find appropriate forwarding state quickly if such state has been established. The addresses do not need to be present if forwarding state is present. In this manner the network protocol syntax, exemplified by the fields contained in the packet header or trailer, can be used for datagrams, virtual circuits or an intermediate form such as soft state-based "flows", where addresses are sent occasionally to refresh forwarding state that might otherwise time out.

5 Flows and Packets

The concept of a network "flow" has been introduced into contemporary discourse. Generally, a network flow is a stream of packets from one source to one or many (in the case of multicast) destinations. Packet streams that require different qualities of service are normally distinguished into different flows even if they have the same source and destination(s); the idea is that packets belong to the same flow if they are to be forwarded along the same path with the same service from the network. Obviously, a virtual circuit is also a network flow. In the Internet, unicast or multicast IP packets from source to destinations have also been termed "flows," especially when these packets contain real-time continuous media like voice and video.

In the next-generation IP protocol, IPv6 [9], there is a new field called the Flow Label that provides at least syntactical support for flows in the Internet. One proposed use for the IPv6 Flow Label is to provide at least one part of a name (along with the source IP address) for an Internet resource reservation made by the RSVP resource reservation protocol [29]. Thus, even in the Internet, one may be able to make reservations on behalf of flows. These reservations are *not* made using classical hard-state based signaling protocols (although they could be). RSVP uses soft-state in hosts and routers to provide flow-based resource reservations from the receiver upstream towards the source of the flow.

The IPv6 flow label is a 24-bit field that is autonomously chosen by the source and is guaranteed to be unique only in combination with the source address. All packets that should receive the same service from the network are labeled with the same flow label by the source. The flow label is meant to simplify classification of IP packets and avoid the need to parse the whole packet to look at port numbers in the transport protocol header. Such parsing can be costly for IPv6 because the header is composed of a list of sub-headers that must be traversed before the transport protocol header can be examined.

The way the flow labels are designed is typical for the Internet architecture; it avoids complexity in the network. Since sources autonomously choose flow labels, no signalling is needed for this purpose. The drawback is that flow labels are not unique unless combined with the source address, and thus cannot be used by themselves to access forwarding information. In virtual circuit networks the VCI is used instead of flow label/source address, but then a signalling protocol is needed to select VCIs.

With some caveats, such as the requirement that the source address must be included with the flow label to individuate a flow, it should be clear that the Internet has borrowed concepts from classical virtual circuit networks to integrate new real time services with traditional Internet services. It is arguable then that IPv6 and RSVP are steps along the way to a hybridization of the Internet. The design of a new network protocol should benefit from the experiences of this hybridization. An aim of a new protocol should be to incorporate both hard and soft-state resource reservation into the network as well as providing support for pure datagram routing.

6   Adapting Headers to Networks

A dominant assumption of packet-switched protocols is that the header (or trailer of a packet) contains fields that are always sent together with each packet. If we look at the fields of a header as separate pieces of control information that can be sent with varying frequencies, we can derive a more flexible network protocol. This principle could also be applied to protocols that are *now* being specified as standards such as IPv6.

For example, as address size grows larger, as in IPv6, or as the need increases to include more than just the source and destination addresses in a packet (for tunneling, etc.), a good use of resources might be to take the large addresses out of some of the packets and substitute a smaller identifier. Soft-state could be established at a downstream switch, with the larger addresses sent downstream periodically to refresh the state. This can provide a new kind of header compression for links in a path that have relatively low-bandwidth. Our design for IPv6 header compression [13], uses this principle to achieve soft-state-based header compression for UDP/IP. On links that have high bandwidth, i.e., where bandwidth is not the dominant factor in the delay of the packet through the switch, the whole header can be reassembled and sent as a unit with the data to provide higher robustness. Indeed, if it is the processing power of the switches that is the forwarding bottleneck, one could add more fields to the header, as suggested in [4], to decrease the processing time in the switch. This is, in effect, the inverse of header compression.

Another example of adapting headers to networks is the use of *header implication* such as can be done when layering IPv6 over ATM. Since the flow label in IPv6 is the same size of the ATM Virtual Circuit Identifier (VCI), it is plausible to map a flow label onto a single VCI. For the common case, when the IPv6 header does not change between the packets in a flow, the IPv6 header need not be sent at all, since it is implied by the ATM VCI and can be recreated on the other side of the IP hop. IP would then present very little protocol overhead from the point of view of the ATM network. This would be especially useful when sending IP traffic over wireless ATM. Moreover, the scheme easily allows quality of service to be preserved over the hop for real time multimedia, if the ATM network provides the quality needed by the IP flow.

7 Switching at one layer only

The Ipsilon flow switching system [18] for switching IP packets in an ATM network is similar to the header implication scheme outlined in the previous section, except that the IP header is not removed. There, a switch that detects an incoming IP network flow[5] that should be switched and forwarded in the same way tells the upstream node to use a separate VCI for packets in that flow. The ATM cells that arrive on that VCI can then be forwarded directly without reassembling the IP packet and examining the IP header. The Ipsilon technique could be used to speed up datagram forwarding over any virtual circuit network technology. For it to work well, it is important that the recognition of network flows is quick and reliable. In the network protocol, the IPv6 Flow Label provides the needed syntactic support.

Ideas similar to Ipsilon's approach for speeding up IP datagram forwarding are now under development in other parts of the Internet community [23, 14, 1, 6]. These techniques are aimed at utilizing ATM switching hardware but do not use ATM connection establishment protocols.

In these schemes, once a packet flow has been detected a VCI is allocated for it and subsequent packets belonging to that flow are sent over that VCI. In this manner the routing lookups and forwarding state lookups need only be done once or at most a few times per packet flow instead of once per packet. Which VCIs are used for what flow is established by a separate protocol.

We believe that the basic idea behind these techniques are general enough to warrant support in the network protocol itself. Having a VCI-like field in network packet headers would allow all switching/routing to occur in one layer only. In this manner, the overall complexity of the network architecture would be reduced. The combination of datagram and virtual-circuit ideas, as exemplified above with flow and tag switching, provides the simplicity and flexibility of datagrams and the fast forwarding capability of virtual circuits.

A new network protocol should recognize this and provide syntactic and signaling support for virtual circuit switching and datagram routing. A new protocol should be able to do one or the other, or a combination of both, as the packet travels from source to destination. This implies that the packets of the new network protocol should be able to be fragmented and reassembled in the network; something that IPv6 forbids. In this way, the new protocol can be used over a variety of network hardware elements, and avoids the necessity of network protocol layering, as seen in IP/ATM architectures with all of their complexities.

8 Hierarchy and Scalability

An important property of a network protocol is its ability to scale to a large number of end systems and network elements. A factor that limits the size of networks

---

[5] The technique can be used even if the IPv6 flow label is zero or if the switch forwards IPv4 packets. Packet streams that are not normally considered flows in the IPv6 sense, for example large file transfers, will also benefit from this technique.

is the amount of routing/forwarding information that routers or switches in the network need to maintain.

For datagram networks, one bottleneck is the size of routing tables. With CIDR [16] a hierarchical structure is imposed on subnet identifiers by assigning subnet identifiers in the same "area" a common prefix. Several routing table entries can thus be collapsed into a single entry consisting of the common prefix. When combined with the "longest-match" strategy for searching routing tables, the original flexibility of the routing table is maintained and the routing topology need not be totally hierarchical.

Virtual circuit network protocols also use hierarchy to decrease the size of forwarding tables. For example, ATM virtual circuit identifiers exhibit a two-level hierarchy. In the backbone of an ATM network, only the first part of the identifier, called the virtual path identifier (VPI), is used when forwarding. Cells that follow the same path through the backbone have the same VPI. This decreases the size of the forwarding table in backbone switches and allows faster switching.

A potential problem with the Flow Labels of IPv6, as currently defined, is that sources pick them autonomously. This makes it difficult to combine several flows that should be forwarded in the same way into a single forwarding entry. If the number of flows passing through a switch is large, the size of the forwarding table can reduce the forwarding speed.

If we consider IPv6, it would be desirable if all flows that are forwarded along the same path and are classified in the same way for quality of service could use the same forwarding information. With the current specification of IPv6 flow labels this is not possible, this is likely to limit the scalability of resource reservation mechanisms in an IPv6 Internet.

9   Reservation Architectures

In traditional connection-oriented virtual circuit protocols, if resource reservations need to be made, they must be made during or after connection establishment. In protocols such as ST-2 or ATM, resources are reserved for real time traffic as the path is being setup in the network, i.e., at connection establishment time. There are some provisions for making or changing reservations during the life of the connection. But there are no provisions for making the reservation in advance of the actual communication. Unfortunately, where network resources are low and time pressures high, e.g., in global teleconferencing where participants reside in widely differing time zones, advance reservations may be necessary for efficient communication. To compare, the airline reservation system would be inefficient for customers and airlines if one could not make reservations in advance.

The case for advance reservations is somewhat better with the proposed resource reservation protocol for the Internet, RSVP. Even though RSVP is designed to provide reservations at the time of communication, reservations in advance of communication can be made. The source of a flow in a multi-party teleconference, for example, sends *PATH messages* downstream periodically through multicast routers to receivers who then send *Reservation messages* upstream towards the source. Reservations for the flow are then made in the routers where Path and Reservation messages meet. If a receiver receives a Path message in advance of a scheduled teleconference, a reservation could then be made and refreshed periodically before as well as during the teleconference. The burden on the network by advance reservations could be minimized by having the sender send Path messages in advance infrequently, increasing the frequency as the time of the teleconference approaches [11].

The problem with this scheme for RSVP is that the sender must be present and sending Path messages for the receiver to be able to make a reservation; and this may be impractical. Also, it should be possible for nodes other than receivers to make advance reservations on behalf of receivers who plan to be part of future communication. A promising approach for providing advance reservations is to see it as an administrative task, i.e., a task for network management. Such a scheme would allow reservations to be made by proxies, without the presence of the actual senders or receivers.

Our conclusion is that the most flexible approach to making resource reservations is to use all three mechanisms mentioned above. When resources are scarce enough so that quality could degrade during a session, but not scarce enough to demand advance reservation, then it may make sense to reserve resources by setting up hard state in a connection establishment phase. When senders and receivers are available before a session starts, soft-state based reservation could be used. Finally, when resources are so scarce as to demand reservations far enough in advance that senders and receivers are not yet in place, it makes sense for network management protocols to administer advance reservations in the network. A reservation architecture for a new network protocol should be prepared to make all three kinds of reservations.

10 Multicast

At least two models of multicast are available in today's communication networks: the group broadcast model used in datagram networks and the point-to-multipoint model used in connection-oriented virtual circuit networks. As usual, each model has its advantages and disadvantages.

The point-to-multipoint model builds a tree from each sender to a group of receivers. This model is exemplified in one version of ATM signaling and in the ST-2 protocol. The model is centralized in that the sender must know the identities of all the receivers, and new receivers can only join the tree by consent of the sender. This centralized approach provides the most security since the sender has almost total control (with the exception of network tapping) over who receives data packets. A disadvantage of this centralized approach, however, is that since the sender must keep track of every receiver, the number of receivers per-sender is limited.

The group broadcast model of multicast, used in Multicast IP [7], builds trees from senders to receivers without the sender necessarily knowing the identities of the receivers. The DVMRP multicast routing protocol for IP, for example, uses flooding and pruning to reach potential receivers on the edges of the network. In addition, local grafting of new nodes onto a pre-existing tree prevents senders from being bothered by changes to the multicast receiver group. The sender does not even have to be a member of the group it is sending to.

On the group broadcast model of multicast, a sender only has to send a packet to a group address and the routers conspire to forward the packets to all the members of the group. This means, however, that a receiver may become a member of a group simply by listening on the multicast address for the group and no consent of the sender is needed. To achieve privacy and security on the broadcast model of multicast, transmissions must be encrypted and receivers may sometimes need to be authenticated at a protocol layer higher than the network.

11 Mobility

An increasing number of users wish to be mobile while accessing the network. This is a problem for some protocols as addresses usually encode topological information, i.e., an address encodes a point of attachment to the network. For example, unicast IP addresses encode a particular subnet and an interface on that subnet.

Better support for mobility is possible if addresses denote a particular end system, regardless of where it happens to be connected to the network. Mobile IP [21, 22] solves the problem by having a *home agent*, located at the mobile host's home network. The home agent captures packets sent to the mobile host and forwards them to where the mobile host happens to be at the time. Forwarding is done by encapsulating packets with an additional IP header whose destination address specifies the mobile's current location. The mobile host reports back to the home agent when it moves again. When the mobile host and its correspondent have established contact, packets can take a shortcut around the home agent and follow the direct route from the correspondent to the mobile host.

In this manner, the address of the mobile host loses its topological significance. The destination address is only used for finding the home agent, after which it is no longer used by the network. There is another kind of IP address that also lacks topological significance: an IP multicast address. At a sufficiently high level of abstraction, the address of the mobile host is like a multicast address that is restricted to having only a single receiver.

The similarity between Multicast and Mobile IP becomes clear when comparing the latter to the PIM multicast routing protocol [8]. In PIM, a *rendez-vouz point, RP,* acts as a place where senders and receivers in a multicast group can meet. Receivers inform the RP that they want packets sent to the multicast group, and senders send their packets to the RP for further delivery to receivers. So, an RP and a home agent have very similar tasks. The principal difference is that there is a single receiver in the Mobile IP case and that authentication of a receiver is vital for Mobile IP but not necessarily for multicast. A network protocol should support both multicast and mobility. It would be surprising if the routing support needed for multicast could not be used to simplify or improve mobility mechanisms.

12   Related Work

Another way to achieve a flexible network protocol is the *ActiveNet* or *Active IP* approach described in [30, 25, 27]. There, packets carry code to be executed by network elements. This provides an easy way to add control or monitoring functions to the network without having to wait for standardization bodies and updating of code in network elements. There are obvious security threats with such schemes and consequently much effort is expended to make the code safe. [27] uses Tcl[2], [25] also mentions Java [17] and various forms of safe platform-dependent binary code.

We agree that the idea of having packets contain the code that network elements should use to process them is very flexible. However, the approach seem to be most useful at moderate speeds; for high speeds, where the processing power of network elements is the bottleneck, it is questionable if such code can be interpreted or compiled and executed fast enough. On low speed networks it is questionable if the increase in packet size can be justified; adding as little as 10 bytes of code to each packet in a flow of interactive audio traffic can increase the required bandwidth beyond what is available.

13   Conclusion

To summarize, a new network protocol should combine the best features of datagram and virtual-circuit packet switching paradigms. To achieve a high degree of flexibility, fields in packet headers can be seen as independent pieces of control information that can be sent with varying frequency in a flow of packets. When a packet does not belong to a flow, all fields can be present. As a packet travels through the network, it might be either switched or routed, or a combination of both, on the path towards the destination(s). Such flexibility allows adaption to various current and future hardware. The new protocol should incorporate both hard and soft-state resource reservation mechanisms, possibly for making reservations in advance. Scalability is an important goal for a network protocol, and thus network addresses as well as flow labels should be hierarchical. Since the routing support for multicast and mobility has some overlap and the problems are similar it might be fruitful to merge the mechanisms.

In this position paper we have tried to lay out some of the fundamental assumptions behind traditional packet switched protocols. We have identified a number of network protocol philosophies such as datagram and virtual circuit models, connectionless and connection-oriented methods of communication, and talked about hybrid models such as network "flows.". In addition we have mentioned different views of resource reservation, hard-state and soft-state, multicast and mobility. We have suggested that a new network protocol incorporate the best features of the foregoing and avoid their pitfalls. We are designing a new protocol, NP++, whose aim is to provide flexibility across many different kinds of networks and hopes to combine the best features of its predecessors.

References

1. F. Baker, Y. Rekhter: *Use of Flow Label for Tag Switching*. Internet Draft (Work in progress), October 8, 1996.
   `draft-baker-flow-label-00.txt`
2. Nathaniel Borenstein: *Email with a Mind of its Own: The Safe-Tcl Language for Enabled Mail*. In Proc. IFIP International Conference, Barcelona, Spain, June, 1994.
3. Robert Braden, Lixia Zhang, Steve Berson, Shai Herzog, Sugih Jamin: *Resource ReSerVation Protocol (RSVP) – Version 1 Functional Specification*. Internet Engineering Task Force, Internet Draft (work in progress), August 13, 1996.
   `draft-ietf-rsvp-spec-13.{txt,ps}`
4. Girish P. Chandranmenon, George Varghese: *Trading Packet Headers for Packet Processing*. Proc. SIGCOMM '95, Computer Communication Review Vol. 25, No. 4, October, 1995, pp. 162–173.
5. David D. Clark: *The Design Philosophy of the DARPA Internet Protocols*. Proc. SIGCOMM '88, Computer Communication Review Vol. 18, No. 4, August, 1988, pp. 106–114. Also in Computer Communication Review Vol. 25, No. 1, January, 1995, pp. 102–111.
6. B. Davie, P. Doolan, J. Lawrence, K. McCloghrie: *Use of Tag Switching With ATM*. Internet Draft (Work in progress), October 8, 1996.
   `draft-davie-tag-switching-atm-00.txt`
7. Steve Deering: *Host Extensions for IP Multicasting*. Request For Comment 1112, August, 1989.
   `ftp://ds.internic.net/rfc/rfc1112.{ps,txt}`
8. Stephen Deering, Deborah Estrin, Dino Farinacci, Van Jacobson, Ching-Gung Liu, Liming Wei: *An Architecture for Wide-Area Multicast Routing*. Proc. SIGCOMM '94, Computer Communication Review Vol. 24, No. 4, October, 1994, pp. 126–135.
9. Steve Deering, Robert Hinden: *Internet Protocol, Version 6 (IPv6) Specification*. Request For Comment 1883, December, 1995.
   `ftp://ds.internic.net/rfc/rfc1883.txt`
10. Mikael Degermark, Mathias Engan, Björn Nordgren, Stephen Pink: *Low-loss TCP/IP Header Compression for Wireless Networks*. To Appear in Proc. MobiCom '96, Rye, New York, November 11-12, 1996.
    `ftp://cdt.luth.se/micke/low-loss-hc.ps.Z`
11. Mikael Degermark, Torsten Köhler, Stephen Pink, Olov Schelén: *Advance Reservations for Predictive Service*. Proc. 5th Workshop on Network and Operating System Support for Digital Audio and Video (NOSSDAV'95), Durham, New Hampshire, April 1995.
12. Mikael Degermark, Björn Nordgren, Stephen Pink: *Header Compression for IPv6*. Internet Engineering Task Force, Internet Draft (work in progress), February, 1996.
    `draft-degermark-ipv6-hc-01.txt`

13. Mikael Degermark, Stephen Pink: *Soft-state Header Compression for Wireless Networks*. Proc. 6th Workshop on Network and Operating System Support for Digital Audio and Video (NOSSDAV'96), Zushi, Japan, May 1996.

14. P. Doolan, B. Davie, D. Katz: *Tag Distribution Protocol*. Internet Draft (Work in progress), September 16, 1996.
draft-doolan-tdp-spec-00.txt

15. David C. Feldmeier: *A Data Labelling Technique for High-Performance Protocol Processing and Its Consequences*. Computer Communications Review (SIGCOMM '93), vol. 23, no. 4, October, 1993, pp. 170–181.

16. V. Fuller, T. Li, J. Yu, K. Varadhan: *Classless Inter-Domain Routing (CIDR): an Address Assignment and Aggregation Strategy*. Request For Comment 1519, September, 1993.

17. James Gosling: *The Java Language Environment: A White Paper*. 1995. Sun Microsystems.

18. Ipsilon, Inc.:*IP Switching: The Intelligence of Routing, the Performance of Switching*. Available from:
http://www.ipsilon.com/productinfo/techwp1.html.

19. Guru Parulkar, Douglas C. Schmidt, Jonathan Turner: *IP/ATM: A Strategy for Integrating IP With ATM*. Proc. SIGCOMM '95, Computer Communication Review Vol. 25, No. 4, October, 1995, pp. 49–58.

20. Vern Paxon: *End-to-end Routing Behaviour in the Internet*. To Appear in Proc. SigComm '96, Stanford University, August 26–30, 1996. A longer version of the paper is available at
ftp://ftp.ee.lbl.gov/papers/routing.SIGCOMM.ps.Z

21. Charlie Perkins, ed: *IP Mobility Support*. Internet Engineering Task Force, Internet Draft (work in progress), May 31, 1996.
draft-ietf-mobileip-protocol-17.txt

22. Charles Perkins, David B. Johnson: *Mobility Support in IPv6*. Internet Engineering Task Force, Internet Draft (work in progress), June 13, 1996.
draft-ietf-mobileip-ipv6-01.txt

23. Y. Rekhter, B. Davie, D. Katz: *Tag Switching Architecture Overview*. Internet Draft (Work in progress), September 17, 1996.
draft-rfced-info-rekhter-00.txt

24. Yakov Rekhter, Tony Li, Editors: *An Architecture for IP Address Allocation with CIDR*. Request For Comment 1518, September, 1993.

25. David L. Tennenhouse, David J. Wetherall: *Towards An Active Network Architecture*. Computer Communication Review, Vol 26, No 2, April, 1996, pp. 5–18.

26. C. Topolcic, Editor: *Experimental Internet Stream Protocol, Version 2 (ST-II)*. Request For Comment 1190, October, 1990.

27. David J. Wetherall, David L. Tennenhouse: *The ACTIVE IP Option*. Proc. 7th ACM SIGOPS European Workshop, Connemara, Ireland, September, 1996.

28. John Wroclawski: *Specification of the Controlled-Load Network Element Service*. Internet Engineering Task Force, Internet Draft (work in progress), August 14, 1996. draft-ietf-intserv-ctrl-load-svc-03.txt

29. Lixia Zhang, Stephen Deering, Deborah Estrin, Scott Shenker, Daniel Zappala: *RSVP: A New Resource ReSerVation Protocol*. IEEE Network Magazine, pp. 8-18, September, 1993.

30. *ActiveNets Home Page*. http://www.tns.lcs.mit.edu/activeware/

Source and Channel Coding for Mobile Multimedia Communications

A.H. Sadka, F. Eryurtlu, A.M. Kondoz

Centre for Satellite Engineering Research
University of Surrey, Guildford, Surrey, GU2 5XH U.K.

Abstract. The transmission of coded video signals is perceptually favoured to provide, to the receiving end, a constant quality with respect to time. In the block type of video coders, like ITU H.263, maintaining a constant quality necessitates a certain amount of consistency in the quantisation process. The step-size quantisation parameter Qp remains unaltered all over the coding process time in order to offer a fixed rate of distortion to the decoded sequence, the fact that is visually perceived as a constant-quality scene. Due to the spatial and temporal tedundancies of raw video signals, maintaining a fixed quantisation parameter leads to the bit rate variability of an encoded video stream. However, the bit rate variability characteristic of an encoded video signal can be considered as an impediment for bit-error detection and correction when a video stream is transmitted over error-prone environments such as mobile radio links, and high interference terrestrial links. This paper incorporates the discussion and implementation of some algorithms applied to ITU H.263 video coding algorithm to enhance its resilience to channel degradation and maintain its performance when its bit stream is transmitted over error-prone media. The resilience techniques are analysed and evaluated through objective and subjective measurements. Moreover, a high quality in-house vocoder for low bit rate applications is described and its error resilience issues are addressed.

1 Introduction

The recent advances in technology have increased the need for communicating video data in highly mobile environments. Services, like mobile videophone, remote video sensing and others, require the implementation of some applications whereby video signals are transmitted in conjunction with speech and data signals over mobile radio links. Integrating video information with other types of data and their transmission over mobile channels define the infrastructure of mobile multimedia communications, which is consuming a crucial share of today's researching efforts.

Unlike leased line and fixed radio channels, mobile environments impose a vast amount of hostility against carried intelligence. Shadowing, multipath fading, various power ranges, radio propagation conditions, interference, coverage constraints and others, are all obstacles that face up the quality of service (QoS) of a transmitted video stream over mobile radio channels. In a fixed radio system, the signal level at the receiving end enjoys almost a guaranteed delivery and correctness, making video error rates extremely low, typically less than 10^{-6}. However, in a mobile environment, channels are more hostile than typical line channels for which video coding algorithms have been designed. In a mobile radio link, error rates can soar up to 10^{-3}, and in some cases as much as 3 to 4%. At these error rates, compressed video bit streams fail to secure an acceptable QoS at the receiving end; therefore, efficient techniques must be devised to protect encoded video data against channel deterioration and improve the resilience of video coders against varying channel conditions.

Coded video streams are indigenously sensitive to error due to the high degree of correlation that exists in its data throughout successive frames. Toggling a single bit in a high activity portion of a frame scope might swamp the ability of the video decoder to correctly recover upcoming data; this is mainly ascribed to the differential type of coding used in the motion estimation and compensation process of a block-type video coding algorithm. The differential coding makes use of the temporal and spatial dependence of video data to achieve a higher compression ratio; consequently, when a bit is flipped, other dependent bits are affected and error accumulates as the decoder is going forward through the video sequence.

In addition to that, another vital factor which affects the resilience of a video coder is the variable bit rate nature of a video bit stream. This property is evidenced by the variable length representation of a 16×16 block of image data. Parameters representing a block are coded with a variable number of bits per parameter, and each parameter might be represented through a variable number of bits in two consecutive blocks. This fact makes it ambiguous to the decoder to locate the start of upcoming data and to identify to what portion of the bit stream it belongs once a single parameter is corrupted on the channel, hence creating a loss of synchronisation that must be recovered before decoder resumes operation.

Unlike video, since speech coding operates on a fixed rate basis, channel hostility problems are not as severe. In-house coders have proved that they could operate with typically 2-3% error rates with very little degradation.

In the following sections, we will describe the major structure of ITU H.263 video coding algorithm, spotting light on the factors that affect its resilience, and consider various ways to make this coder more robust and resilient to channel errors. In addition, the result of performance variation of in-house speech coders under various channel conditions will be presented.

2 ITU-T H.263 Very Low Bit Rate Video Coding Algorithm

International Telecommunications Union (ITU) is considering the study and review of a video coding algorithm, namely H.263, which is intended for very low bit rate coding (less than 64 kbit/s) for PSTN networks , and its standardisation process is in the final stages.

H.263 is a block transform based video coder, and is very similar, in its layering structure, to ITU H.261 video coding algorithm: The picture layer is a set of Group of Blocks (GOB), which is in its turn a set of 16×16 matrices of pixels called Macroblocks (MB), each of which is a group of 6 8×8 blocks, 4 for the luminance component and 2 for the chrominance one. Besides, H.263 uses discrete cosine transform (DCT) and huffman coding methods to encode different parameters in the 2 coding methods : Intra and Inter.

However, the two coders are very different in details and performance. H.263 introduces different symbol representations in order to increase the efficiency of entropy coding. The resulting Variable Length Codewords (VLC) might vary in length for two consecutive MBs. If a bit is toggled in the first MB, the decoder will be unable to figure out the length of the corrupted symbol in order to skip it and flush its buffer to resume decoding the next VLC. The most trivial solution for the decoder to recover its synchronisation is to search for the first error-free synch word, which is located at the beginning of each sequence frame. This phenomenon is one major issue that characterises the resilience dilemmas encountered in an H.263 video bit stream when it is transmitted over mobile channels.

In H.263 video coder, each Interframe coded Macroblock (MB) consists of a motion vector (MV) and a residual matrix which is the difference between the currently coded MB and the best matching one (in the previous frame) pointed at by the associating MV. In the motion estimation process, H.263 video coding algorithm compares the currently processed MB against MBs in the previous frame within a user-defined search window sliding on all directions of the MB position. A sum of absolute differences (SAD) is calculated between the current MB and each of the MBs that lie inside the search window. The MB which scores the least SAD is considered to most match the current MB, and the relative SAD is the MB residual matrix. the resulting SAD is matched against a threshold value to decide whether Intra or Inter mode is to be used for a specific MB. If Inter frame is selected, H.263 encoder passes the coefficients of the difference matrix through DCT, zigzag-pattern, run-length and huffman encoders respectively, and starts its motion compensation process which is depicted in Figure 1.

The motion compensation process in H.263 shows the dependence of MBs on previously transmitted information to calculate the candidate predictors of their MVs. This steps up the degree of correlation amongst the video data bits, causes the effect of error on one MV bit to swiftly propagate in both time and space, and gives rise to a loss of synchronisation at the receiving end, the fact which gets the decoder to skip all upcoming bits until it locates the first intact synch word to retrieve its synchronisation.

On the other hand, more accurate interframe prediction is achieved in H.263 coder by employing half-pixel accuracy motion estimation. This cancels out the need for a spatial filter used in the motion compensation loop of H.261 to mitigate the effect of blocking artifacts, and doubles the picture dimensions by linear interpolation.

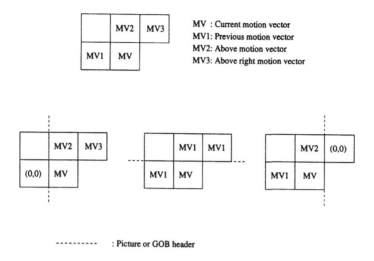

Fig. 1. Motion compensation of H.263 encoder

The most prominent development in H.263 involves the negotiable options which can be switched on or off depending on the input signal. Obviously, an agreement between the encoder and decoder is necessary as some decoders may not support all of the options. There are 4 negotiable options:

Unrestricted motion vector

In this mode, the pixels located outside the frame are predicted from the border pixels and used in motion compensation. This increases the subjective quality when there is motion near the frame borders, and also may reduce the overall bit rate as much as 10%.

Syntax-based arithmetic coding

In this option, syntax-based arithmetic coding is used for entropy coding of parameters. This allows a bit rate reduction dependent on the symbols sequence, but obviously increases the computational complexity of the codec.

Advanced prediction mode

When motion information of one macroblock containing 16×16 luminance pixels cannot be represented accurately by a single vector, this option can be switched on for that particular macroblock and a separate motion vector can be used for each block of 8×8 pixels. Although there are more motion vectors to be encoded, the subjective quality can be significantly improved especially when there are small moving objects in the sequence.

PB frames

This option involves bidirectional motion compensation in a way similar to the MPEG standards. It allows doubling of the frame rate with a slight increase in bit rate. On the other hand, if the frame rate is kept constant, the image quality may be improved or the bit rate may be reduced.

From the coder resilience point of view, when PB frame option is on, the encoded bit stream looks more susceptible to a damage in QoS. Since the delta motion vectors of a B-frame depend on the motion vectors of the 2 P-frames from which it is predicted, a corrupted bit would facilitate the propagation of error more rapidly and on a wider scope. In other words, an error in a P-frame MV would lead to errors in at most 3 other MVs in the next P-frame due to the motion compensation process, and 2 other delta vectors, 1 in each of the next and previous B-frame. This will speed up the propagation of error throughout the sequence frames and cause the accumulation of errors much more widespread in the picture area.

Figures 2 and 3 show the subjective and objective effects of channel errors on an H.263 video bit stream encoded at 64 kb/s with all negotiable options switched off.

Fig. 2. 200th frame of H.263 encoded "Foreman" sequence at 64 Kb/s with negotiable options all off when transmitted on : (a) error-free channel (b) error-prone channel of BER=10^{-4}

3 Error Resilience Techniques

H.263 video coding algorithm has been designed to provide very low bit rate compression of video signals for transmission over fixed telephone networks (PSTN). The aspects of error control and robustness have not been specifically addressed. Yet, the resilience of the video coder needs to be improved so that H.263 can be employed to provide mobile services. Efforts to develop error resilience techniques are being exerted in parallel with the development of MPEG-4 video coder. These proposed techniques constitute the core experiments for error resilience issues of MPEG-4 video coder[3]. The effect of a mobile channel can very well be simulated by a random error generator since the bursty error effect of a mobile link on a slice (GOB or frame) of a video bit stream is very much similar to that of a single bit error. This is due to the fact that when synchronisation

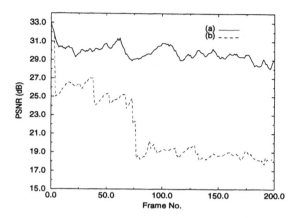

Fig. 3. PSNR values for "Foreman" sequence encoded at 64 Kb/s with negotiable options all off when transmitted on : (a) error-free channel (b) error-prone channel of $BER=10^{-4}$

is lost, H.263 skips all the bits of the slice, regardless of their correctness, until the next error-free SW is located. This is equivalent to saying that the skipped bits are corrupted by a bursty occurrence of error. In this section, we discuss 3 possible error resilience techniques applied on an H.263 encoded bit stream to counteract the effect of mobile channel hostility on QoS and synchronisation. To assess the performance of those techniques and their efficiency, we have set our target bit rate at 64 Kb/s so that we can assess the practicality of the error resilient coder in real life. By setting a target bit rate, we achieve a fixed coding efficiency which might be degraded due to the inclusion of the protection bits in the video data bit stream. Therefore, the error resilient coder is compared to the original H.263 coder operating at the same bit rate. This comparison will highlight the effectiveness of the error resilience technique and its effect on the subjective quality of the video signal without taking into consideration the degradation resulting from a coarser quantisation step size. On the other hand, objective measurement using PSNR graphs is not always a proper way of assessing the effectiveness of an erroneous video bit stream. Due to loss of synchronisation, frames might split or merge resulting in a sequence disorder. A frame split at the beginning of a sequence implies that all the forthcoming frames will rely on the wrong reference frames for the calculation of PSNR values. This sequence disorder will remain until a frame merge occurs and sequential order is recovered.

3.1 Duplicate Information

To limit the accumulation of errors in both temporal and spatial domains, the probability of a MV data bit being corrupted should be minimised. The resilience of a video bit stream can thus be greatly improved by transmitting the MV information twice at different locations of the bit stream. Hence, the probability of receiving an erroneous MV data bit is reduced. To enable the video decoder to locate the duplicate information in the variable rate bit stream, a synch code, different from the picture start synch code (PSC), is used. On the other hand, motion vectors are not the only piece of data that this error resilience technique might consider to duplicate. Intra DC coefficients, MB Modes are other highly critical information which contribute to the enhancement of resilience. Figure 4 depicts the layering structure of H.263 bit stream when the duplicate information technique is applied on a MB level.

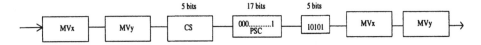

Fig. 4. H.263 coder MV duplicate information applied on a MB level

Due to the variable length of the x and y components of MV data codewords, 2 kinds of error might arise : 1 or more bits are flipped in a VLC word making the decoder unable to map it to any of the valid VLCs available in the MV tables of VLCs. In this case, the decoder being aware of the occurrence of an error, flushes its buffer searching for the duplicate information to replace the erroneous VLC. The second sort of error, however, is more ambiguous: One or more bits are toggled in a way that the resulting erroneous VLC matches one the valid VLCs in the decoder MV tables. One possible way to avoid letting this error occur without being detected is to transmit a checksum (CS) representing the parity bits for each (x,y) components of a MV. If no error is dug out after calculating the parity and comparing it to CS, the decoder assumes an error-free MV data, and skips all duplicate information after locating its start through the appearance of the first synch word (SW). To differentiate between this SW and PSC, SW is followed by a particular bit pattern of 5 bits (21). Figure 5 shows the subjective improvement achieved by this technique. However, a major drawback of this technique is the massive number of bits it adds to the stream making it undesirable for almost most low bit rate video coding applications. A total of 27 bits are transmitted for each Inter coded MB together with the MV duplicate information.

Additionally, this technique is susceptible to absolute failure when SW itself is affected by the channel hostility. In this case, the decoder would skip it, looking for the first intact SW which precedes the MV duplicate information of another MB.

In order to reduce the huge number of redundancy bits used in the above mechanism, MV duplicate information is sent once each GOB. However, this will impose a higher delay on the decoding process since the decoder has to freeze its operation waiting for all the GOB data to arrive until it can locate the SW word and read the duplicated version of the affected MV data. If the channel

190

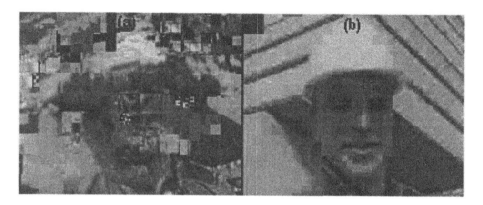

Fig. 5. 125th frame of H.263 encoded "Foreman" sequence with negotiable options all off with MV bit stream transmitted over an AWGN channel of SNR = 2.0 dB : (a) normal H.263 bit stream (b) H.263 bit stream with duplicated MV information on a MB-level

error rate is considerably high (in the range of 0.01% and up), the jerkiness becomes noticeable and very much annoying to visual perception. To protect SW against channel errors and lessen the probability of failure of this technique, an RS(5,7) code was used for the synch word. This increases employed overhead, but enables the channel decoder to detect and correct 1 bit-error in the synch word before the bit stream is presented to the H.263 video decoder. The total overhead imposed on this technique is solely attributed to the RS coded SW word of 21-bit length. This will result in an overhead of 4.8 Kb/s for a frame rate of 25 f/s. Figure 6 depicts the layering structure of H.263 bit stream when the duplicate informaion is applied on a MB-level.

Figure 7 shows the improvement on the subjective quality achieved by sending the GOB MV data twice in a bit stream while keeping a constant bit rate. Figure 8 shows an objective measurement of this resilience technique. Due to a frame split, the sequence order is lost and decoded frames are no more associated with their reference frames in the original sequence within the calculation of PSNR values. For that reason, we see that the PSNR graph of the duplicate data video signal goes below the normal bit stream one for more than 60 frames before a frame merge occurs. When the frame order is retrieved, the graph of the error resilient bit stream starts diverging from the original H.263 bit stream graph.

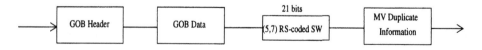

Fig. 6. MV duplicate information applied on a GOB level in H.263 video coder

Fig. 7. 199th frame of H.263 encoded "Foreman" bit stream with negotiable options all off, encoded at 64 kb/s and transmitted over a channel of BER = 10^{-5} (a) normal H.263 bit stream (b) H.263 bit stream with duplicated MV information on a GOB-level

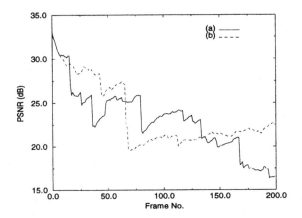

Fig. 8. PSNR values for "Foreman" sequence encoded at 64 Kb/s with negotiable options all off and transmitted over a channel of BER = 10^{-5} (a) normal H.263 bit stream (b) H.263 with duplicated MV information on a GOB-level

3.2 FEC Channel Protection Techniques

To combat the effect of highly erroneous channels encountered in mobile environments, the use of FEC schemes proved to be an efficient choice to maintain the resilience of speech and data coders. However, for video encoded signals, the main discrepancy that urges a change in the way FEC techniques are applied, is that they enjoy a variable bit rate characteristic.

A protected variable length codeword generates another VLC after FEC techniques are applied. This makes it impossible for the channel decoder to identify what portion of the bit stream has been protected, and leads to complete confusion as to what segment of the data belongs to what parameter of the video bit stream. Consequently, in our FEC experiments, our main concern was to tackle this problem caused by the variability of bit length in the video parameters. We have protected the MV bit stream against channel errors using a 1/2 rate convolutional channel coder with a constraint length of 7. Since the maximum number of bits that any MV x or y component in H.263 video coder can occupy is 13, we have assumed each input codeword to the convolutional channel coder to be of length 13. If it happens that the MV component bit length is less than the maximum, we pad it up with extra bits in order to input it to the channel coder as a fixed length codeword. The channel decoder, in its turn, decodes fixed-length segments of the bit stream before forwarding the corrected data to H.263 video decoder. The performance of this FEC technique with the padding-up scheme applied to the 1/2 rate convolutional coder is best illustrated in Figures 9 and 10.

Fig. 9. 100th frame of H.263 encoded "Carphone" bit stream with negotiable options all off, encoded at 64 kb/s with MV bit stream transmitted over an AWGN channel of SNR = 2.5 dB : (a) normal H.263 bit stream (b) H.263 bit stream with MV bit stream protected using a 1/2 rate convolutional coder (Overall BER = 0.57%).

3.3   Two-way decoding and Reversible VLC

As the name implies, the 2 way decoding makes use of the reversible VLCs which are readable in two directions to help locate and confine the affected area of a bit stream. If an error is detected in the forward direction, the decoding process is immediately interrupted and the next synch. code is searched. Once SW is located, the decoding process is resumed in the reverse direction. This scheme enables the decoder to start decoding at two different locations of the bit stream without the need to send additional information such as synch words. The combination of reversible VLCs and the two-way decoding are very effective at combatting the effects of both random and bursty errors.

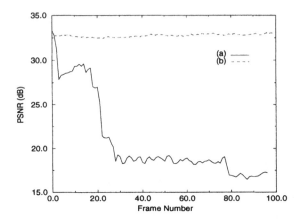

Fig. 10. PSNR values of "Carphone" sequence encoded at 64 Kb/s and sent over a channel of SNR = 2.5 dB (a) normal H.263 bit stream BER = 0.57% (b) MV codewords protected with a 1/2 rate convolutional coder

When both the forward and reverse direction decoding processes are terminated, the decoder follows one of four approaches to decide what portion of the bit stream is to be discarded. These approaches are depicted in Figure 11.

In (a), the data between the point where an error is detected in the forward decoding process and that detected in the backward decoding process is discarded. Some information may still be available to the decoder for these corrupted MBs. In (b), the MBs whose data is free from errors is acceptable. data surrounded by the points where errors occur in the forward and reverse directions is discarded. In (c), Macroblocks, where data is corrupted, are dropped and not used in the decoding process. In (d), only the corrupted MB is discarded and its data is not used in the decoding process.

To generate a RVLC, either one of many techniques can be used. One approach could be to adopt a predetermined number of 1's and use it in any particular codeword. For example, if three 1's are included in one code, a possible RVLC would be : (111, 1011, 1101, 11001, 10101, ...). Another methodology to generate a RVLC might assume the same number of 0's and 1's in each codeword. In this case, a possible RVLC would be : (01, 10, 0011, 1100, 001011, 000111, 110100, ...).

In our simulations, we have used an RVLC table whose words contain a fixed number of a symbol (1). An error-free first Intra frame is assumed and a random error generator is used to simulate the mobile channel. Although mobile channels impose a bursty type of hostility on carried intelligence, the effect of an error burst on a portion of a video bit stream is quite similar to that of a random error. When a segment of a bit stream is affected by error in one-way decoding, resulting in a state of desynchronisation, the whole segment of data is skipped until the decoder locates the first upcoming error-free SW; this will simulate the

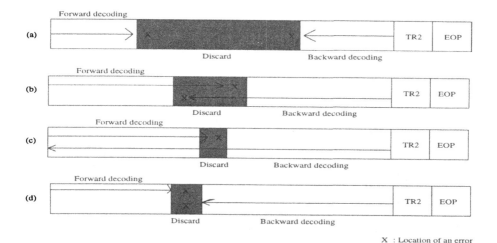

Fig. 11. Categories used in 2-way decoders to determine what portion of the bit stream to discard : (a) Separated detected error points (b) Crossed detected error points (c) Error is detected in only one direction (d) Errors isolated to a single MB

effect of loss to a burst of errors striking all the skipped bits regardless of their correctness and validity (Figure 12).

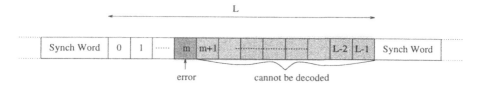

Fig. 12. One-way decoding of variable length codes

Therefore, a single bit error in video bit streams usually has bursty error effects. To enable the fast recovery of synchronisation, a GOB start code was used at the beginning of each GOB. However, in two-way decoding, the occurrence of an error does not indicate the loss of the relative GOB since backward decoding can identify the correctly received bits which were skipped in one-way decoding due to loss of synchronisation (Figure 13).

Figures 14 and 15 show the subjective and objective improvements achieved on a video signal by applying the two-way decoding algorithm on video streams RVLC codewords.

Since a single bit error in video bit streams usually implies bursty error effects, the effective bit error rate (BER) may be much higher than the channel BER. Assuming that there are L bits in a GOB (between two consecutive correctly detected synch words) and the channel BER is e_{ch}, then the effective BER e_{eff}

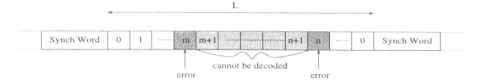

Fig. 13. Two-way decoding of variable length codes

Fig. 14. Carphone sequence encoded at 28 Kb/s and subject to one bit error for : (a) one-way decoding (b) two-way decodable reversible VLC codewords

can be derived as follows:

$$e_{eff} = \sum_{m=0}^{L-1} p_m \frac{L-m}{L} \qquad (1)$$

where p_m is the probability of having the first corrupted bit at the bit position m as illustrated in Fig. 12, and it is a function of e_{ch} and m:

$$p_m = (1 - e_{ch})^m e_{ch} \qquad (2)$$

Then, Equation (1) can be rewritten as:

$$e_{eff} = \frac{e_{ch}}{L} \sum_{m=0}^{L-1} (1 - e_{ch})^m (L - m) \qquad (3)$$

Equation (3) demonstrates that the effective error rate is actually much larger than the channel error rate in usual one-way decoding.

The block length L is an important parameter which affects the relation between effective and channel BERs. Due to the variable rate nature of a video signal, the block length is a variable depending on the encoder parameters such as the quantisation parameter and negotiable options. However, the average block

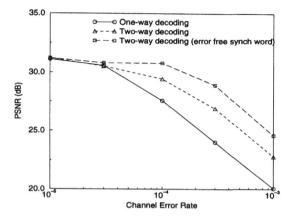

Fig. 15. PSNR values for Carphone sequence encoded at 28 Kb/s using original and two-way decodable codes

length can be estimated by using the overall bit rate r, frame rate f, number of synch words per frame s and the length of the synch word l_s :

$$L_{ave} = \frac{r}{fs} - l_s \tag{4}$$

In the case of two-way decoding, when an error is detected in forward decoding, the decoder can find the next synch word and start decoding in the reverse direction. In this way, more information can be correctly decoded. With reference to Figure 13, the effective BER for two-way decoding can be calculated as follows:

$$e_{eff} = \sum_{m=0}^{L-1}(1-e_{ch})^m e_{ch}\left((1-e_{ch})^{L-m-1}\frac{1}{L} + \sum_{n=0}^{L-m-2}(1-e_{ch})^n e_{ch}\frac{L-m-n}{L}\right) \tag{5}$$

$$e_{eff} = (1-e_{ch})^{L-1}e_{ch} + \frac{e_{ch}^2}{L}\sum_{m=0}^{L-1}\sum_{n=0}^{L-m-2}(1-e_{ch})^{m+n}(L-(m+n)) \tag{6}$$

Figure 16 shows that the effective BER for two-way decoding converges to the channel rate at low BER values while there is always an off-set between them in the case of one-way decoding.

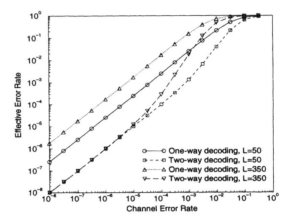

Fig. 16. Effective error rate in one- and two-way decoding as a function of channel error rate and block length

4 Speech Coders Resilience

An in-house speech coder operating at low bit rates is used. The coder which is described in this paper is an improvement to previous MB-LPC vocoders [4][5] in terms of synthetic speech quality and robustness to background acoustic noise. This algorithm may be operated over a wide range of bit rates, from as low as 1.8 Kb/s up to 4 Kb/s depending on the quantisation schemes and frame update rate employed. Figure 17 shows a schematic of the encoder which shows the bit allocation for a 24 Kb/s version using a 20 ms frame length.

Input speech is pre-processed using a high-pass filter to remove any DC component. A tenth order Durbin's algorithm is then applied to find the spectral envelope which is quantised in the LSF domain using a multi-stage split vector quantiser. The quantisers parameters are then used to find the LPC residual signal whose spectrum is approximately flat. Pitch is determined from the original speech using a new modified sinusoidal model matching algorithm, based upon the technique proposed by McAulay[6]. This produces highly reliable pitch information, even when presented with acoustically noisy speech. Pitch refinement is then applied, which attempts to match a voiced synthetic spectrum to the original spectrum using a method akin to MBE [7]. the synthetic spectral error over each harmonic band is used to make a binary voicing decision for each harmonic. By examining the voicing distribution throughout the 4kHz spectrum, a single voicing frequency is determined above which the excitation is assumed to be unvoiced, and below which it is voiced. finally, the amplitudes of all harmonic are calculated and are used to determine the levels of a fixed number of spectral bands, which are then vector quantised.

Figure 18 depicts a schematic of the decoder. Once the parameters have been decoded from the bit stream, the harmonic amplitudes are modified to reduce

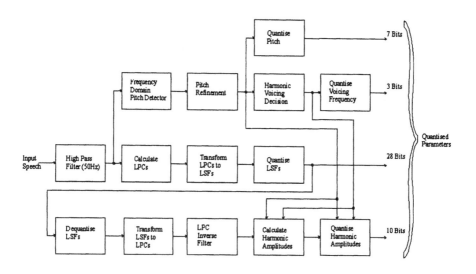

Fig. 17. Speech encoder schematic

the energy levels in the valleys of the LPC spectral envelope, thus improving the quality of the synthetic speech. The unvoiced excitation is generated by spectrally shaping a spectrally flat noise source. the voiced excitation takes the form of a new sinusoidal harmonic generator which produces a more natural sounding synthetic output than the MBE algorithm [7] and requires fewer instruction cycles to implement. The unvoiced and voiced excitation is summed and filtered using an LPC filter whose coefficients are updated every 5ms by applying linear interpolation in the LSF domain. This produces the final synthetic speech output.

Initial studies indicate that at a rate of 2.4 Kb/s using the bit allocation shown in Figure 17, the new MB-LPC vocoder produces synthetic speech at a higher quality than the new U.S. Federal Standard 2.4 Kb/s MELP coder [8]. The algorithm shows very little degradation in speech quality as the bit rate is lowered to 2.0 Kb/s. With more frequent frame update however, the coder's performance is increased, at a bit rate of 4.0 Kb/s it produces equivalent quality to an 8 Kb/s CELP coder. When tested with background acoustic noise, even though the synthetic noise is modified, the synthetic speech quality is better than that produced by an equivalent CELP coder. To make this vocoder robust to channel degradations, the pitch needs to be protected against channel errors; yet if it is lost, we can replace it by the previous pitch value. Other information is error resilient : the voice/unvoice values in case of an error detection. When amplitudes are lost, they are all set to 1.

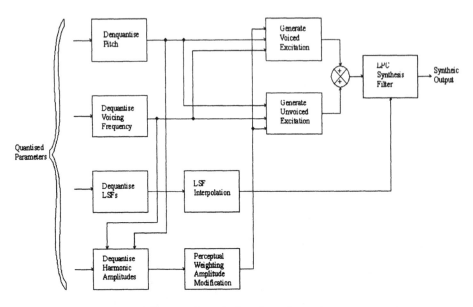

Fig. 18. Speech decoder schematic

5 Conclusions

Video signals are extremely sensitive to errors due to their variable bit rate characteristic and the high correlation amongst their parameters both in time and space. Consequently, to transmit them over high BER channels such as mobile radio links, a mechanism must be applied to the video stream to protect its bits against channel corruption, maintain the resilience of the video coder, and retrieve its synchronisation which might be lost as a result of a single bit flip. The mechanisms, presented in this paper, gave promising results that could be adopted to enhance the resilience of ITU-T H.263 video coding algorithm against channel hostility in mobile environments. The duplicate information mechanism gave a reasonable bit protection efficiency when applied on a GOB level and the redundancy appeared to be quite feasible in this case. FEC techniques, such as convolutional and RS codes are quite useful to protect a specific parameter in the bit stream by padding it up with bits belonging to subsequent parameters in the bit stream, to act against the variable bit length nature of the video VLC codewords. Two-way decoding helps in locating the erroneous bits and spotting the affected area to minimise the effect of error on the overall QoS and recover the synchronisation of the video decoder. Finally, the in-house speech coder proved to outperform in quality the new U.S. Federal Standard 2.4 kbit/s MELP coder[8] and proved to be more robust to channel degradations with BERs going as down as 2×10^{-3}.

6 References

[1] Draft ITU-T Recommendation H.263 "Video Coding for low bit rate communication", November 1995.

[2] Telenor R&D "H.263 Video codec test model", November 1995.

[3] International Organisation for Standardisation, ISO/IEC JTC1/SC29/WG11. Description of Error Resilient Core Experiments, Document No: N1327, July 1996.

[4] S. Yeldener, A. M. Kondoz, B. G. Evans "Multi-Band Linear Predictive Speech Coding at Very Low Bit Rates", IEE Proc. VSSP, October 1994, Vol. 141 No. 5, pages 289-295.

[5] S. Yeldener, A. M. Kondoz, B. G. Evans, "Perceptually Based Modelling of Excitation Spectrum for the Applications of Multi-Band LPC Speech Coding", Proc. Int. Conf. on Telecoms, Istanbul 1996, Vol. 1 pages 148-151.

[6] R. J. McAulay, T. F. Quatieri, "Pitch Estimation and Voicing Detection Based on a Sinusoidal Speech Model", ICASSP, Albuquerque 1990, pages 249-252.

[7] D. W. Griffin, J. S. Lim, "Multi-Band Excitation Vocoder", IEEE Trans. ASSP, Aug. 1988, Vol. 36, No. 8, pages 1223-1235.

[8] A. McCree, K. Truong, E. B. George, T. P. Barnwell, V. Viswanathan, "A 2.4 kbit/sec MELP Coder Candidate for the New U.S. Federal Standard", IEEE Trans. ASSP, Atlanta 1996, pages 200-203.

Developing a Conference Application on Top of an Advanced Signalling Infrastructure

R.J. Huis in 't Veld [a], A-N. Ladhani[b], F. Moelaert El-Hadidy[c],
J.P.C. Verhoosel[c], B. van der Waaij[b], and I.A. Widya[b]

[a] Philips Multimedia Business Networks[1], P.O. Box 32, 1200 JD Hilversum, The Netherlands.
[b] University of Twente-CTIT, P.O. Box 217, 7500 AE Enschede, The Netherlands.
[c] Telematics Research Centre, P.O. Box 589, 7500 AN Enschede. The Netherlands.

Abstract

Within the Dutch project Platinum[2] an advanced signalling infrastructure has been developed and implemented. This infrastructure is based on the results of the RACE MAGIC project. It allows users to start up and manage multimedia calls between multiple parties and can be configured to the specific needs of users.

Platinum also defines and implements a framework of component applications on top of this infrastructure. The strength of this framework is its flexibility to combine component applications into groupware applications satisfying the specific needs of a group of end-users.

In this paper, we present a part of the software architecture of this framework. It addresses audio and video exchange, conference management, floor-control of media, templates containing the roles of users, and retrieval of information necessary for displaying the human computer interfaces. The paper also demonstrates the flexibility of the advanced signalling infrastructure.

1. Introduction

Today, much work is put into the development of flexible platforms [Alt+93,Alt+96,SaBe95] allowing for the development of a wide variety of groupware applications that satisfies the needs of end-users. These platforms of tomorrow will have to abstract from the heterogeneous infrastructure as much as possible, possess trading or brokering functionality to link service users to service suppliers, and have generic components applications that can be combined into groupware applications satisfying the specific needs of end-users.

The Dutch project Platinum contributes to this field by developing an advanced B-ISDN signalling infrastructure together with generic component applications running on top that can be integrated into groupware applications in, for instance, the fields of tele-education and tele-consulting. Conceptually, this infrastructure has three layers (see Fig. 1). The lowest layer is the ATM layer with AAL-5 and SAAL on top. The

---

[1] The author participated in Platinum while he was working for the University of Twente.

[2] The partners in the Platinum project are Lucent Technologies, the Telematics Research Centre, and the University of Twente-CTIT. The Platinum project is funded by the Dutch Ministry of Economic Affairs.

second layer extends primarily the standard ATM control plane signalling with a multicast (one to many). The protocol used for this is Q.2931ext and is based upon the results of the RACE II project MAGIC-R2044 [Sme+94]. The third layer is called the middleware. This layer is also split into a control plane and a user plane. The control plane handles the signalling for setting up and managing multimedia, multiparty calls using the user data fields in the service primitives of the underlying signalling plane. The user plane transforms multimedia streams into bit streams before they are transported over AAL-5 connections and, vice versa, the bit streams coming from AAL-5 connections into multimedia streams that are offered to e.g. end-users. The middleware is configurable to satisfy the specific needs of its user.

The component applications running on top of the middleware are grouped into: conference management, shared whiteboard, and structure editor. Conference management sets up and manages the groupware application such that audio, video, and data can be exchanged between multiple parties. It also supports templates for setting up conferences of various types, floor control for each of the media used in a conference, and firewall capabilities that regulate the roles and rights of end-users. The second group provides end-users with a shared whiteboard on which they can draw, change, and delete objects. This component application is based on a shared object medium where all objects are replicated at each of the participants in the shared whiteboard meeting. A single floor-controller controls the access to these objects. The third group of applications provides the end-users with the support of collaborative editing on documents using, for instance, stand-alone wordprocessors. These structure editor applications are based on the facilities found in the tool-set of opendoc that use the ORBs of CORBA [OMG91]. These ORBs interact using the same type of shared object medium that is developed for the shared whiteboard application. For all three groups of applications, Human-Computer Interfaces (HCIs) were also developed.

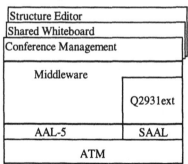

Fig. 1. Layered architecture of Platinum infrastructure.

The Platinum system distinguishes itself from other systems as it uses the B-ISDN approach to integrate ATM signalling with multiparty multimedia call signalling. Other projects such as BERKOM [Alt+93,Alt+96], MICE[SaBe95], and Beteus [Blu+95] use computer network protocols like TCP/IP protocol suite or IP next generation as a starting point to abstract from the underlying infrastructure. A second difference is that the Platinum middleware provides a flexible environment to set up multimedia calls defined in a library and to add new types of multimedia calls to this library. The other systems have only dedicated solutions for each type of multimedia call. A third difference is that the Platinum system is in principle telecommunication based. That is, anybody should be able to request a conference at the moment of need (like using a telephone) and when the resources are available the conference is provided. The other systems use a reservation or registration system instead. Finally, the component applications in Platinum are not stand-alone. They are embedded in the middleware and have HCIs with a common look and feel.

A software architecture has been defined for all three groups of applications as well as for the provisioning of the HCIs. These software architectures have been implemented using Visual C^{++}4.0. The authors of this paper defined the software architecture [Tee+96] for the conference management application and the HCI provisioning. The approach followed in determining this software architecture is based on the vertical approach advocated in [OuSc95], meaning that both user requirements as well as technical constraints of the middleware [Dri95] were taken into account. The user requirements were derived by setting the pilot use of the conference application to tele-education [Lad+95]. For this area, main actors and their roles were identified and task analysis for each of these actor roles was performed, displaying key functionality as required by the actors to fulfil their goals. To generalise the task analysis outcomes to other areas than tele-education, they were used to select a set of general user requirements for interactive systems from a checklist with 109 such requirements.

In this paper, we present the software architecture of the conference management application and the HCI-provider as they are embedded in the middleware. In section 2, we start by presenting the software architecture of the nucleus of the middleware: the middleware kernel. We also show how it can be extended. Section 4 presents the software architecture of the conference management application. In section 5, the software architecture of the HCI provider is presented shortly. Finally, we draw some conclusions in section 6.

2. Middleware Kernel

The Platinum network consists of an ATM network enhanced with Q.2931ext; i.e. Q.2931 extended with multiparty calls. On top of this network, a layer is defined via which users (be it participants or applications) can start up and manage multiparty, multimedia calls. This layer is called the middleware (Fig. 1) and is configurable to the specific wishes of users. It consists of a fixed part called the middleware kernel and of user-defined extensions. In this section, we present the software architecture of the middleware kernel at a local end-point, i.e. the terminal equipment. This software architecture is split into three parts: session model, control, and view. We will discuss each of them shortly. This section is concluded by demonstrating how the middleware kernel can be extended.

2.1 Session Model

A session model is associated with each user of the middleware kernel. This session model contains information locally available about the calls in which that user partakes. A session model of an user corresponds to an instance of the class diagram in Fig. 2 (class diagram in Booch notation [Boo94]). It shows that during a session a user can participate in zero or more calls. It also shows that there should at least be two parties in each call. When these parties interact using a medium (e.g. audio, video), the medium is represented by a medium object in the model and each party is related to that medium object via a PartyMediumEdge.

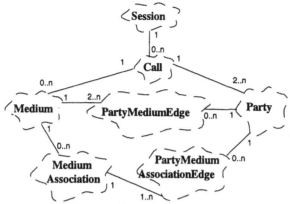

Fig. 2. Class diagram of session model.

Currently, the Platinum network provides unidirectional and bi-directional network as well as multicast (one to many) connections. Moreover, a medium may be transported over the network using several of these connections (e.g. separate connection for audio and video). So, it is possible that a party needs several network connections to exchange information. This is reflected in the session model by having each of these connections represented by a medium association object linked to the parties involved via a PartyMediumAssociationEdge.

Attributes are associated with each object in the session model. To allow the middleware to be configurable, each object has user-defined attributes which can be used to maintain user-specific information about a call. Furthermore, a browser is available to retrieve information from the session model.

2.2 Control

The control part of the middleware kernel provides users with two generic operations *Add* and *Delete* to affect a session. For instance, via the Add it is possible to start a new call or add a party or medium to an existing call.

The sequence diagrams of the Add and Delete show the primitives that can be exchanged remotely when these operations are invoked (Fig. 3). Notice that after an AddReq, participants can get an AddInd to which they respond with an AddRspAck. Then, the network evaluates all these responses and sends an AddCnfAck to the initiating participant and an AddCmpAck to all the others. Notice that the Delete is not negotiable. Everybody gets informed that a delete will take place, and only the initiating participant is informed a delete has taken place successfully.

Users of the middleware kernel will have to have references to middleware objects generating these primitives in order to receive them. Furthermore, A user has to create a session object in this way before adding it to the session model. Session objects are created using the well-known factory mechanism.

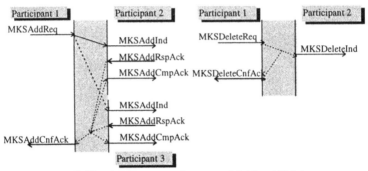

Fig. 3. The time sequence diagrams of Add and Delete.

2.3 View

The view part (or view as it is often called) processes the information received via the network connections to, for example, information visible on the screen or coming out of the speakers. Conversely, the view processes information on the desktop before it is placed on the network connection. It consists of stacks of objects, called transporters, that handle, amongst others, compression, synchronisation, and (de)multiplexing of streams. The view reflects the session model. When control changes the session model, the session model initiates the appropriate changes that have to be made to the view.

Stacks are created by the MediumBuilder that uses a library containing all kinds of stack configurations. This library can be altered off-line by application developers, allowing so for flexibility. Furthermore, application objects that wish to interact with the view objects need to have references to them.

2.4 Extension

The middleware kernel is extendible. In the current version, the modification of attributes in objects of the session model is not supported via the call processing. However, in order to function properly, our conference application definitely needs to modify user-defined attributes in session objects. Therefore, we extend the middleware kernel with a mechanism for global modification of attributes in session objects that functions outside the call processing.

This mechanism extends the view with a selective multicast data medium stack via which modify messages can be exchanged with an extended session convener. For efficiency reasons, a single modify multicast medium exists for each call and is created during the set-up of that call. The session convener gets a reference to this medium and handles all incoming messages. If a Modify message is received, it informs the control to make the necessary changes to the session model. Indirectly, this will then change the view. All parties in the call have send and receive access to the modify medium. A two-phase commit protocol is used to guarantee that the versions of the session object at all sites involved are identical

3. Conference Application

In this session, we outline the functionality required of our conference application. The discussion is split into three parts: conference management, conference configuration, and floor control.

3.1 Conference Management

The study of the user requirements (see section 1) suggested that the organisational structure of a Platinum application should consist of a conference, participants, groupmeetings, and media. A conference denotes the overall environment allowing users to exchange information. A participant is somebody who partakes in a conference. A groupmeeting (e.g. lecture) is a self-contained group of participants in a conference that exchange multimedia information. There may be multiple groupmeetings within a conference. A participant (e.g. student or teacher) can only exist in a groupmeeting. Finally, a medium is an association between participants in a groupmeeting to exchange information. A medium can vary from an audio, video, or data stream connection to a more compound form such as shared whiteboard or compound structure editor.

Conference management offers participants the possibility to set up and maintain several groupmeetings within the conference. The set-up can be done by hand, specifying each individual component in the conference, or by using an already predefined set-up (template). After the set-up, each participant will get his or her own view of the conference. Such a view can be a global construction of the complete conference, or a local view containing only the relevant information for the participant: the media and participants he is directly associated with.

Each medium can be individually controlled, the sound level is adjustable for audio media, the control of objects on the shared-whiteboard can be altered, etc. Beside these changes, each medium can be attached or detach dynamically during the life-cycle of the groupmeeting among a (sub)-set of participants within the groupmeeting. This mechanism is called floor control. Furthermore, it is possible to modify existing groupmeetings by adding, removing or modifying participants, media and associations between participants and a medium. Moreover, new groupmeetings can be added by participants already present in a groupmeeting.

3.2 Conference Configuration

Platinum has to provide a conference that is configurable to the wishes of participants. Therefore, templates have been added to the conference functionality. Templates are used to outline the type of conference and thereby the responsibility of each participant in the conference and the resources that are available to him or her. For instance, within Platinum we associate a manager with each conference, groupmeeting, and medium. A conference manager is unique to a conference and is by default the participant who sets up the conference. The same applies to the groupmeeting manager. A medium manager is only defined if a floor control mechanism is being used to gain access to a medium. Furthermore, for each groupmeeting a participant is in, a role is defined. This role defines participant's rights to carry out certain actions within a groupmeeting. We distinguish rights related to that groupmeeting (e.g. Add_Participant) and rights related to a medium function (e.g. Grab_Floor). A medium function defines how a medium is used. For instance, in the field of tele-education, an audio medium can be used in a question-and-answer medium function in a lecture groupmeeting as well as a chat medium function in a student assignment groupmeeting.

In order to be able to define a specific conference configuration, we use two types of templates: style template and conference template. The style template specifies the layout of a specific conference without participants. It contains the following items:

- Definition of different groupmeeting styles offered by a conference. In a groupmeeting style the collection of media to be used are identified. The choice to use the floor control mechanism (or not) is specified for each medium. Then a collection of different roles that a participant can take are defined in terms of management rights and medium access rights.
- Layout of a conference is defined by choosing one or more groupmeeting styles belonging to the same or different type.

The conference template is the style template populated with the actual participants. It is used as a parameter during the set-up of a conference. It consists of the following items:

- A database of participants that may participate in the conference.
- A list for a particular conference is generated using a specific Style Template. For the tele-education scenario this may be a lecture about ergonomics given to the second year students. This list contains the following:
    ◊ List of the participants that are invited to take part in the conference.
    ◊ The association of participants with the roles they have in each groupmeeting they participate in.
    ◊ If a medium uses a floor control mechanism then one participant has to be assigned as the floor controller of this medium.

3.3 Floor Control

In our conferencing application, a set of participants is associated with a medium. Each of these participants has the right to read from the medium, the right to write to the medium, or both. In most application domains, it is desirable to control the access of the participants to a medium. This is known as floor control. In this context, floor control means temporarily disabling a participant to read from a medium or temporarily disabling a participant to write to a medium. In this paper however, we only consider the latter one.

Whenever a participant adds a medium to a groupmeeting, he or she can indicate the need for floor control. If no floor control is needed, each participant associated to the medium can always access the medium according to his access rights. Otherwise, the following floor control policy will be used:

- There is a single floor controller which is allowed to enable and disable access to the floor. For now, we also assume that the role of floor controller cannot be transferred to another participant.
- There is a maximum number of floor holders which have access to the floor. The application ensures that this maximum is not exceeded.
- The following floor control functionality is provided for each floor:
  ◊ a request for the floor to the floor controller can be done by all participants, except the floor controller,
  ◊ a grant of the floor can be done by the floor controller, either as a response to a request for the floor or on the floor controller's own initiative,
  ◊ a reject of a request for the floor can be done by the floor controller,
  ◊ a grab of the floor from a current floor holder can be done by the floor controller on his own initiative,
  ◊ a leave of the floor can be done by a current floor holder on his own initiative, and
  ◊ a withdraw of an outstanding request for the floor can be done by a participant that currently has an outstanding request for the floor.

4. Software Architecture

In this session, we outline the software architecture of our conference application. Likewise section 3, the discussion is split into three parts: conference management, conference configuration, and floor control.

4.1 Conference Management

Fig. 4 shows the software architecture of conference management. The HCI (Human-Computer Interface) and the middleware are drawn in grey. The information streams from the middleware to the HCI are represented by the big grey arrow. The management commands are represented by the black arrows. Between the HCI and the middleware lies the conference management (CM) system. This system has three parts: The CM Provider which is the access point for the HCI to the CM-system. The CM Notify part which handles the notification messages from the middleware about changes in the Session model and forwards these to the HCI. The third part is the CM Model containing the state information about the current conference.

The first responsibility of the CM Provider is to check if a participant initiating certain commands via his HCI has the proper rights for this. This is done by asking the CM Model what rights this participant has. These rights are then used to check if the command can be executed or that it must be refused. Secondly, the CM Provider must translate the CM commands into commands understandable to the middleware and send them to the middleware (the command arrow).

After the middleware has processed these commands, it will update its session model. Each change in this model will be reported to the CM Notify who translates and forwards the notifications from the middleware to the HCI. The HCI is responsible for updating the screen to inform the user of the changes. The second task of CM Notify is to keep the CM Model up to date with the MW Session model. This can be done by updating the CM Model at each useful notification from the middleware.

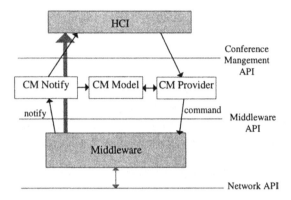

Fig. 4. The software architecture for conference management.

The middleware session model contains information about multiparty, multimedia calls. Additional information is necessary to transform a session of multiparty multimedia calls into a conference with groupmeetings and participants. This information is stored in the CM-Model. Fig. 5 shows the class diagram of the CM-Model. Here we see that a CM-model consists of one conference object and zero or more groupmeeting objects. Zero or more participants objects are associated with each groupmeeting object, and each participant object can occur in multiple groupmeetings. Furthermore, there is an edge object to capture the role that a participant plays in a groupmeeting. This is in line with organisational structure of a Platinum application (see section 3.1).

The attributes of the objects in the CM-model ensure that a conference object is related to the session object in the session model of the middleware and that there exists a 1-1 relation between groupmeeting objects and call objects. As the middleware only allows a party object to be associated with one call object, the List Party Ids attribute in a participant object outlines the party objects associated with it. Other attributes are added to maintain information about who is controlling the conference and the groupmeetings, and what role a participant plays in a groupmeeting.

Medium objects are not referred to in the CM Model because the middleware handles them in exactly the same way as the CM provider wants it. There is no extra information necessary. The only extension of the media is floor control. This is treated entirely as an extension of the middleware, and therefore cannot be found in the CM Model.

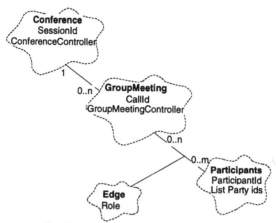

Fig. 5. Class diagram of the CM-model.

4.2 Conference Configuration

Fig. 6 shows the object oriented description of the software architecture for conference configuration. The StyleTemplate database and the ConferenceTemplate database are relational databases that may reside in memory or in files on the harddisk. They contain the style templates and conference templates respectively. When a user needs to use the databases, the HCI creates the TemplateMaker, TemplateFiller and TemplateModelManager. The TemplateModelManager knows the structure and location of the ConferenceTemplate and StyleTemplate databases and is responsible for the creation and modification of both databases. The HCI uses the TemplateMaker to add, delete, and modify the StyleTemplate database. The TemplateMaker uses the TemplateModelManager to carry out these changes. The TemplateFiller is used by the HCI to add, delete, and modify the objects in the ConferenceTemplate database. The TemplateFiller in turn uses the TemplateModelManager to implement these changes in the ConferenceTemplate database.

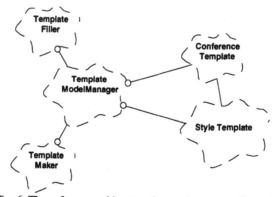

Fig. 6. The software architecture for conference configuration.

When a conference is started, the HCI starts up the objects as described above and informs the CM Provider of the TemplateModelManager identity. The CM Provider can then access all information about the templates via the TemplateModelManager. In the case of setting up a conference, the CM Provider gets a message set-up from the HCI with a reference to the conference template object to be used. It then fetches the information by browsing the database via the TemplateModelManager and generates the appropriate commands to the middleware.

4.3 Floor Control

Floor control is realised using the modify multicast medium and the generic modify scenario that were defined as an extension of the middleware (see section 2.4). Furthermore, we assume a reliable delivery of messages at the destination. The classes relevant for the software architecture of floor control are depicted in Fig. 7.

When a medium is added, the Medium object in the middleware session model gets two user-defined attributes: FloorControllerID, that indicates the participant who is floor controller, and MaxFloorHolders, that indicates the maximum number of participants that can hold the floor simultaneously.

When a participant is added to a medium, the PartyMediumEdge object in the middleware session model gets an application-defined attribute FloorStatus, that indicates the current status of floor with respect to the particular participant. This status can either be:

1. HOLD, when the participant holds the floor,
2. REQUEST, when the participant has an outstanding request for the floor, and
3. NONE, when the participant does not have the floor and does not have a request pending.

In addition, the PartyMediumAssociationEdge object (associated with an outgoing network connection) gets an user-defined attribute AttachedWrite, that indicates whether the participant's outgoing stack is currently attached or detached. All floor control actions are carried out by modifying the user-defined floor control attributes using the generic modify scenario.

In the RequestFloor scenario, the ServiceController (see Fig. 7) at requester's site performs the generic modify scenario to change the FloorStatus attribute in the PartyMediumEdge of the local participant and involved medium to the new value REQUEST. When this scenario succeeds, the HCI at the floor controller site is notified and informs the floor controller that a new request has arrived.

In the GrantFloor scenario, the ServiceController checks whether the number of PartyMediumEdge objects with FloorStatus set to HOLD (the number of floor holders) is less than the MaxFloorHolders attribute in the Medium object. If so, the ServiceController performs the generic modify scenario (1) to change the FloorStatus attribute of the PartyMediumEdge of the granted participant and the involved medium to the new value HOLD, and (2) to change the AttachedWrite attribute of the PartyMediumAssociationEdge of the granted participant and the involved outgoing medium association to the new value SET. When this scenario succeeds, the HCI at the non-floor holders site is notified and informs the participant that the floor is granted to him. In addition, the MediumBuilder at the non-floor holders site is notified and attaches the outgoing stack.

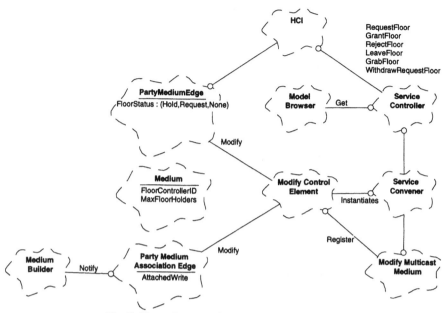

Fig. 7. The software architecture for floor control.

In the RejectFloor scenario, the ServiceController performs the generic modify scenario to change the FloorStatus attribute of the PartyMediumEdge of the rejected participant and the involved medium to the new value NONE. When this scenario succeeds, the HCI at the requesters site is notified and informs the participant that the request for the floor is rejected.

In the GrabFloor scenario, the ServiceController performs the generic modify scenario (1) to change the FloorStatus attribute of the PartyMediumEdge of the grabbed participant and the involved medium to the new value NONE, and (2) to change the AttachedWrite attribute of the PartyMediumAssociationEdge of the grabbed participant and the involved outgoing medium association to the new value RESET. When, this scenario succeeds, the HCI at the floor holders site is notified and informs the participant that the floor has been grabbed from him. Similary, the floor controller is informed that the grab has been succesful. In addition, the MediumBuilder at the floor holders site is notified and detaches the outgoing stack.

In the LeaveFloor scenario, the ServiceController performs the generic modify scenario (1) to change the FloorStatus attribute of the PartyMediumEdge of the local participant and the involved medium to the new value NONE, and (2) to change the AttachedWrite attribute of the PartyMediumAssociationEdge of the local participant and the involved outgoing medium association to the new value RESET. When this scenario succeeds, the HCI at the floor controller site is notified and informs the floor controller that a floor holder has left the floor. In addition, the MediumBuilder at the local (floor holders) site is notified and detaches the outgoing stack.

In the WithdrawRequestFloor scenario, the ServiceController performs the generic modify scenario to change the FloorStatus attribute of the PartyMediumEdge of the local participant and involved medium to the new value NONE. When this scenario succeeds, the HCI at the floor controller site is notified and informs the floor controller that a request has been withdrawn.

5. HCI Provisioning

A Human-Computer Interface was designed by taking into account the user requirements as well as technical constraints resulting from the underlying middleware software. They were balanced in solving the design problem. A design solution was chosen in which the HCI was made pluggable so that different HCI styles and views for different (human) user types and task context could be easily placed and exchanged on top of the middleware software. Pluggability is achieved using a software architecture based on Microsoft Visual C++ MFC. (see Fig. 8).

Consistent with the CM provider and the middleware, we divide the HCI provisioning into two parts: (1) the presentation of the media, and (2) the presentation and control of the status of the conference. The presentation and interaction with the media are self-contained. They are called media views and are based on the software drivers of the audio and video cards used. The presentation and control of the status of the conference is constructed out of three parts. The HCI View, responsible for the interaction with the user. This includes the presentation of the status and the capturing of user commands. The second part is the HCI Data, a database containing all the information necessary to keep track of the status of the conference. For instance, HCI Data contains information relating CM-provider commands to the appropriate GUIs of the user. The last part is the HCI Doc, it co-ordinates the flows between the HCI View, the CM provider and the HCI Data.

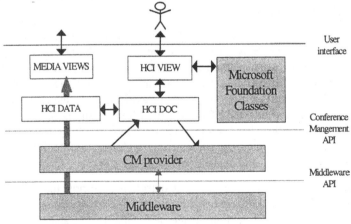

Fig. 8. The software architecture for making the HCI pluggable.

6. Conclusions

This paper reports the implementation-oriented high-level design of a conference management, a floor control and the integrating human computer interface (HCI) components. These components form a part of a groupware application framework that collaborates and co-exists on top of a true B-ISDN-based telecommunication infrastructure, which uses control-plane signalling to provide multimedia, multiparty call support. The designed components are described in terms of their functionality and are elaborated in the software architecture.

The applied design approach is basically a vertical approach. It takes user requirements into account but allows design choices to be constrained by the underlying telecommunication infrastructure (in particular by the middleware part of the infrastructure). In-line with this approach, the developed HCI integrates the interfaces of the components of the groupware application to provide a common look and feel to the application end-users and to discriminate and constrain the usage of the application in accordance with the role of the user in the particular application.

In the first version developed, the groupware application is based on a tele-education setting that applies an instructional didactic model. This, for example, is reflected in the characteristic of the floor control which does not allow passing of the floor controller role to another groupmeeting participant. However, more generic user requirements have also been considered resulting in the introduction of Conference Templates, Style Templates and groupmeeting objects that contain modifiable participant roles, medium management and medium access rights. These objects enable extensions to support generic groupware applications.

During the design process, the software architectures of the HCI, floor control, and the conference management are chosen to be similar to the architecture of the middleware. This simplifies the usage relations between the objects in the HCI, the conference management, floor control and the middleware. Also the relation with the shared whiteboard application and the structure editor application have been taken care of. This alignment has consumed significant project manpower for example for extensive e-mail discussions and regular joint meetings (often weekly) between the different taskforces working in the various areas. The alignment effort was a complex task because of the parallel development of all parts of the application and the middleware. However, the effort has paid off in the implementation phase. The openness of the middleware, which enables functionality extensions easily, and the high abstraction level of the middleware functionality narrow the design gap between the middleware and the facilities needed by groupware applications. That is, the middleware provides a sufficient large basic functionality that often only need to be configured at the application layer.

The middleware may for example be extended by a modify operation on session objects in the call processing (i.e. control plane) part. This extension does not only simplify floor control implementation but it also makes the infrastructure more generic. With this extension and together with a further development of the defined Style and Conference templates general groupware applications can be supported, for example computer supported co-operative work applications and tele-education applications based on other didactic models than the instructional model previously addressed.

A follow-up to the Platinum project is called MESH (Multimedia services on the Electronic Super Highway). This project is funded by the Dutch Ministry of Economic Affairs and aims at accelerating the introduction of advanced, relevant and flexible telematics services for the Electronic Super Highway. Thereby, the Platinum technology will be applied in user-oriented pilots in the field of tele-education, tele-consulting, tele-cooperation and tele-meeting. This concerns real-life situations in the Dutch educational and medical sector. Further extensions and adaptation of Platinum applications for the Electronic Superhighway will be made by studying and implementing facilities necessary for pilot settings, and eventually for commercial use.

7. References

[Alt+93]   M. Altenhofen et al. The BERKOM Multimedia Collaboration Service. ACM multimedia'93, 1993.

[Alt+96]   M. Altenhofen et al. The BERKOM Multimedia Teleservices, Volume II: Multimedia Collaboration, Release 3.3.1, January 1996.

[Blu+95]   C. Blum et al. A Semi-Distributed Platform for the Support of CSCW Applications, First International Distributed Conference on High Performance Networking for Teleteaching (IDC '95), Madeira Portugal, November 1995.

[Boo94]   G. Booch. Object-oriented Analysis and Design with Applications. 1994.

[Dri95]   B van Driel editor. Middleware Software Architecture, Platinum/T2.1/N035 /V06, November 1995.

[Lad+95]   A-N. Ladhani, I. P. F. De Diana, and I.A. Widya. Combining Desktop Tele-classroom and Distributed Multimedia Databases for Tele-tutoring, First International Distributed Conference on High Performance Networking for Teleteaching (IDC '95), Madeira Portugal, November 1995.

[OMG91]   OMG. The Common Object Broker Request: Architecture and Specification, OMG TC Document Number 91.12.1, Revision 1.1, December, 1991.

[OuSc95]   H. Ouibrahim and J. Schot. Tele-teaching and the electronic superhighway: Towards a vertical approach, First International Distributed Conference on High Performance Networking for Teleteaching (IDC '95), Madeira Portugal, November 1995.

[SaBe95]   M.A. Sasse and R. Benett. Multimedia Conferencing over the Internet, The MICE Project. Library&Information Systems Briefings, issue58, March 1995.

[Sme+94]   A. De Smedt, P. Hellemans, L. Ronchetti, and K. Täubig. A Multilevel Framework for the Static Description of Telecommunication Services, Signaling Protocols and Services for Broadband ATM Networks 7 (2), 1994.

[Tee+96]   W. Teeuw, R.J. Huis in 't Veld, L. Ferreira Pires, and H. Bakker editors. Specification and design of Platinum applications, Conference Management, Shared Whiteboard, and Collaborative Structure Editing, Platinum/T2.3/N034/ V00, February 1996.

New Network and ATM Adaptation Layers for Real-Time Multimedia Applications: A Performance Study Based on Psychophysics

Xavier Garcia Adanez, Olivier Verscheure and Jean-Pierre Hubaux

Telecommunication Services Group - TCOM Laboratory
Swiss Federal Institute of Technology
CH-1015 Lausanne, Switzerland
E-Mail:{garcia, verscheure}@tcom.epfl.ch
URL: http://tcomwww.epfl.ch/{~garcia, ~verscheu}

Abstract. We present in this paper Network and ATM Adaptation Layers for real-time multimedia applications. These layers provide a robust transmission by applying per-cell sequence numbering combined with a selective Forward Error Correction (FEC) mechanism based on Burst Erasure codes. We compare their performance against a transmission over AAL5 by simulating the transport of an MPEG-2 sequence over an ATM network. Performance is measured in terms of Cell Loss Ratio (CLR) and user perceived quality. The proposed layers achieve an improvement on the cell loss figures obtained for AAL5 of about one order of magnitude under the same traffic conditions. To evaluate the impact of cell losses at the application level, we apply a perceptual quality measure to the decoded MPEG-2 sequences. From a perceptual point of view, the proposed AAL achieves a graceful quality degradation compared to AAL5 which shows a critical CLR value beyond which quality drops very fast. The application of a selective FEC achieves an even smoother image quality degradation with a small overhead.

1 Introduction

ATM technology is reaching a certain level of maturity that allows for its deployment in local as well as in wide area networks. Concurrently, audiovisual applications are foreseen as one of the major users of such broadband networks. However, it has already been shown in [1, 2] that the cell and frame loss ratios might not be negligible in ATM based environments especially if the operators employ statistical multiplexing to efficiently use network resources. MPEG-2, the standard for full motion video, will be the audio and video compression tool of such multimedia applications. The semantic and syntactic structure of the data flows generated by MPEG-2 makes the applications sensitive to data loss. Errors can spread across a single image causing holes or may spread out over several images within a Group of Pictures (GOP). The adoption of AAL5 by the ATM Forum [3] for the transmission of MPEG-2 video even increases the impact of data loss due to its packet level granularity which implies packet discard in case of cell loss. To reduce the impact of cell losses on video applications, using a cell level granularity reduces the CLR seen by the application compared to AAL5 based transmission [4].

Traditional error recovery methods are based on retransmission, also known as Backward Error Correction (BEC). While this technique has proven to be very efficient for data applications, it fails to protect real-time multimedia applications due to their stringent timing constraints, especially in Wide Area Networks (WAN) and point-to-multipoint configurations. Conversely, Forward Error Correction (FEC) techniques do not rely on data retransmission but instead, by adding redundancy, are able to correct the errors at the receivers end. The major drawback of FEC techniques is that they add delay and overhead. We propose in this paper a Burst Erasure [5] selective FEC technique based on the analysis of the syntactic and semantic components of the stream to be transmitted. We efficiently protect the most sensitive elements of the data by adding a small overhead since the header to raw data ratio is small. ATM Adaptation Layers (AAL) by definition are generic in the sense that they have to support any kind of applications, albeit not any ATM Transfer Capability (ATC). To get rid of any application specificity we have developed a Network Adaptation specific to MPEG-2 on top of the AAL. This layer is able to identify the headers encapsulated into the Transport Stream (TS) packets [6]. We show by simulation that the proposed AAL gives better results in terms of cell loss. As it is very difficult to map the impact of cell losses onto MPEG-2 based video applications, we use a perceptual quality metric based on psychophysics to bring to the fore the improvements obtained by our proposal from the end user perspective [7].

The paper is organized as follows: Section 2 describes the requirements of today's multimedia applications and the currently available mechanisms. In the next section we develop the mechanisms that we propose for a multimedia AAL and we describe the operation of the FEC in combination with a Network Adaptation. Section 4 describes the simulation setup and shows the improvements measured in terms of network and perceptual quality parameters. We derive some conclusions in Sec. 5 and we provide an outlook of how this work will be continued.

2 ATM Adaptation Layer Mechanisms for Multimedia Applications

2.1 Requirements of Multimedia Applications

The principal characteristic of multimedia data streams is that it is of a continuous nature. As such, it has very stringent constraints in terms of delay and delay jitter. This has been referred to as Timely Information [8]. This means that data arriving beyond a certain point in time is considered as lost by the application. Moreover, if the audiovisual information makes use of compression techniques such as MPEG-2, it becomes also sensitive to loss due to the lack

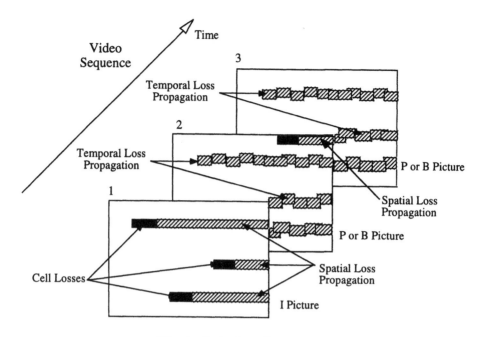

Fig. 1. *Data Loss Propagation*

of redundancy in the transmitted data albeit the impact of loss on the image heavily depends on the type and the location of the lost information. Data loss spreads within a single picture up to the next resynchronization point (e.g. slice headers). This is referred to as spatial propagation. Due to the predictive nature of the MPEG-2 algorithm, when losses occur in a reference picture (Intra-coded, I, or predictive, P, frames) it will remain until the next intra-coded picture is received. This causes the errors to propagate across several pictures which is known as temporal propagation (see Fig 1). The impact that the loss of syntactic data may have is in general more important and difficult to recover than the loss of semantic information. So, the transport of real-time multimedia data has to be reliable and timely. ATM networks fulfill both conditions but, in some cases, they may fail to guarantee loss ratios.

2.2 Current AAL Mechanisms

Nowadays, two AALs are available for the transport of multimedia data: AAL1 and AAL5 [9]. AAL1 was basically designed to cover circuit emulation services. It therefore offers Constant Bit Rate (CBR) services to the applications. This may be a limitation if constant quality encoding or Variable Bit Rate (VBR) in general is used. The advantages of AAL1 are that it provides a cell level granularity to detect cell losses via a sequence number and also provides cell loss recovery via FEC combined with interleaving. However, using interleaving introduces delays proportional to the size of a FEC block at both the sender and the receiver. The FEC scheme applied is based on Reed-Solomon codes that are able to correct erasures (cell losses) and also random errors (impulse noise). However, the bit error ratios to be considered in fiber-based ATM networks are close to 10^{-12}. In fact, with such impulse noise error values, it becomes unnecessary to have bit or octet granularity. Therefore, and taking into account that it introduces considerable delay and overhead, an octet based interleaver is not a major need.

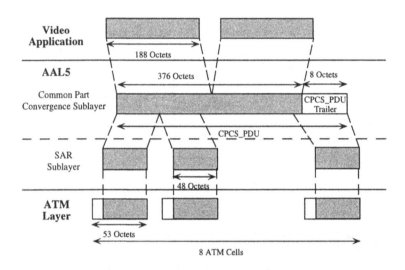

Fig. 2. *AAL5 Segmentation Mechanism of MPEG-2 Video*

On the other side, AAL5 is able to cope with any ATM transfer capability. One of the reasons for this is its simplicity and low overhead. It receives PDUs from the upper layer, appends an 8 byte trailer containing a CRC-32 parity check, a length indicator and padding information for boundary alignment, and sends the AAL5-PDU to the Segmentation and Reassembly (SAR) sublayer. This simplicity has however some drawbacks for multimedia applications. AAL5 does not provide enough protection against erasures because it was mainly designed for data transfer applications that rely on robust transport protocols for error correction. It provides an error detection mechanism based on the parity check calculated on a PDU basis and the check of the length of the received packet. Due to the lack of more sophisticated error detection functionalities, in case of cell losses, i.e. length indicator mismatch, AAL5 is unable to know the position of the cells lost inside the PDU and so, no error correction even at a higher layer can be applied. Therefore, when cell losses are detected, the packets are discarded. If no retransmission mechanism is applied, the packet discard leads to a data loss at the application level which is higher than the data loss (cell loss) at the network level. This is clearly not adequate for real-time applications for two main reasons. First, a single cell loss causes the loss of several data that could be still used by the decoder with techniques such as early resynchronization [10]. Second, AAL5 does not notify errors to the upper layers. Therefore, the decoder may not detect data corruption and may not be able to apply any of the multiple error concealment mechanisms [11].

3 New Mechanisms for Real-Time Multimedia Applications

3.1 ATM Adaptation Layer Mechanisms

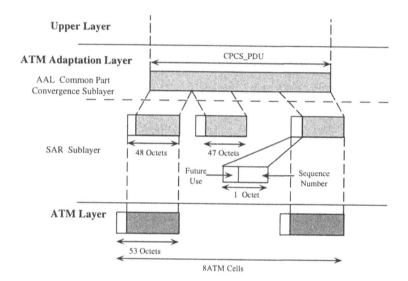

Fig. 3. *Proposed AAL Segmentation Mechanism*

To meet the requirements of real-time multimedia applications, we propose an AAL that includes the following mechanisms: cell level granularity to improve the error detection capability and a selective FEC mechanism to selectively protect essential data. The cell level granularity needs the introduction of a per-cell sequence number. To make this possible, we use 47-byte cell payloads which frees an octet per ATM cell to insert the sequence number and also control information for the FEC mechanism (see Fig 3). If we consider the transport of MPEG-2 TS packets, our packetization has the advantage of always giving an integer number of cells, since the length of the TS packets is 188 bytes which gives exactly 4×47 byte payloads. This increases the per cell overhead. However, we have reduced the PDU overhead since we do not add any header or trailer information as AAL5 does. Since our packetization process does not provide such kind of PDU delineation, our approach consists in assuming that the PDU size is negotiated at connection setup and that it will remain constant for the duration of the connection. Therefore, the receiver does not need any extra information to delineate the packets. Moreover, this scheme reduces the number of states that the protocol has to deal with in case of loss. This scheme allows to reduce the data loss seen by the receiver because it increases the resolution of the data loss detection algorithm avoiding the packet discard. The receiver uses the sequence numbers to detect the number and position of the lost cell in a packet. This information can thus be passed to the upper layers that can take necessary action to conceal data loss. In our proposal, we use a dummy cell insertion mechanism, when cell losses are detected (see Fig 4). This has two advantages: first, it guarantees packet length integrity. Second, in the case of MPEG-2, it can be used as an error message since sequences of 47 zero-octets are not allowed by the MPEG-2 standard [12]. So this mechanism can be exploited by the decoder as an error indicator. Also, if FEC is used, the insertion of dummy cells simplifies the error correction. However, since the proposed mechanisms are not specific to MPEG-2, we also consider the fact that dummy cell insertion may be useless or even harmful to other applications. Consequently, we propose this mechanism to be user selectable at connection setup.

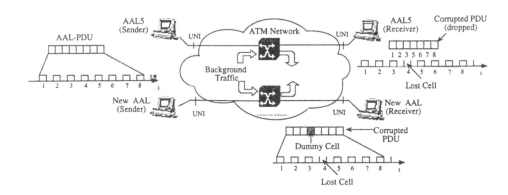

Fig. 4. *Dummy Cell Insertion Mechanism*

The utilization of FEC for real-time multimedia applications has several advantages. Besides the fact that it introduces relatively low delay compared to retransmission based error correction mechanisms, it is also better suited to point-to-multipoint configurations that should be widely used by such multimedia applications. The major drawback of FEC is that it introduces overhead. Indeed, it adds redundancy packets that will be able to recover the missing data when losses are detected; moreover, FEC does not guarantee zero loss. To reduce the overhead generated by the FEC scheme, we propose to include a selective mechanism in our AAL. The advantage of such a selective method is twofold: first it reduces the overhead, and second, it adapts to highly structured information such as compressed video. Indeed, the impact of syntactic data loss (i.e. headers) is rather different than the loss of semantic data. Headers are points of synchronization. When a header is lost, all the underlying data cannot be recovered. This leads to a loss of quality and also to a waste of network resources since data correctly received could not be decoded which may either generate artifacts or loss of synchronization. Moreover, such selective mechanism can be used to protect separately audio from video since losses in audio are much more noticeable and disturbing to the user than video losses. Besides, also, timing information can be selectively protected to reduce the probability of losing synchronization.

The flexibility proposed by our mechanism cannot be achieved by using the method already developed for AAL1. The latter uses Reed-Solomon Codes with interleaving which gives a fixed matrix structure. Due to this structure it is difficult to apply a selective mechanism. We propose a mechanism based on burst erasure codes (RSE) [5]. The advantage of this method is that it takes into account the specifics of ATM. It relies on the fact that erasures are limited to fixed boundaries (cells) to correct cell losses only. Conversely, it cannot correct octet or bit errors (impulse noise). This is not a fundamental problem since we assume that the ratio of errors to erasures is very small and can be considered as negligible.

3.2 Network Adaptation Mechanisms

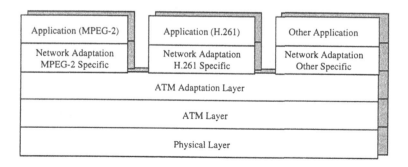

Fig. 5. *Protocol Stack*

By definition, an AAL has to be generic in the sense that it does not have to provide mechanisms specific to applications. The mechanisms proposed in Sec. 3.1 albeit designed with real-time applications in mind are not specific to any audiovisual coding scheme. To provide efficient protection of data with selective FEC mechanisms, it is necessary to know the syntax of the data to be transmitted. The protocol stack depicted in Fig. 5 presents the network adaptation as the layer specific to an application. In our example, we have used MPEG-2 based applications so we have developed for our experiments a network adaptation layer specific to this standard. The network adaptation main functionalities are: delivery of fixed size PDUs (in the case of MPEG-2 it is straightforward) which includes packet segmentation/reassembly and alignment to boundaries (padding), and detection of loss sensitive data combined with the generation of FEC_request messages. The second functionality in the case of MPEG-2 applications consists of detecting headers and generating FEC_request messages accordingly. Based on these messages passed with the PDU, the AAL will or will not generate the FEC data.

The proposed Network Adaptation depicted in Fig. 6 receives the TS packets from the MPEG-2 system layer and generates a single NA-PDU. Since we assume a fixed packet size, no PDU delineation is done and therefore no overhead is generated. However, applications using other encoding systems such as JPEG or MJPEG that generate variable length packets will need to add some information to align the packet sizes to PDU boundaries.

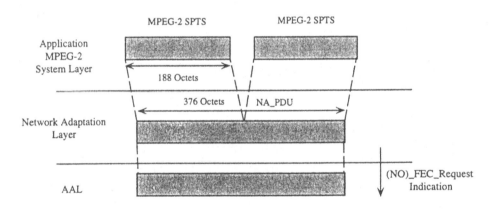

Fig. 6. *Proposed Network Adaptation for MPEG-2 based applications*

4 Comparison of AAL5 and the Proposed AAL

4.1 Simulation Setup

The simulation setup used for our experiments is depicted in Fig. 7. The simulator is composed of four multimedia workstations and two ATM switches. Both switching stages, implemented as multiplexers with limited buffer size, are loaded with background traffic provided by several On-Off sources. This type of

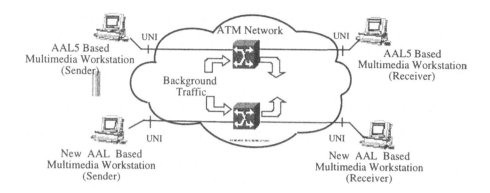

Fig. 7. *Simulation Scenario*

source model is widely used to simulate a multiplex of traffic such as the one that could be found at the entrance of an ATM switch. Moreover, two state Markov source models encompass the peak cell rate parameter which is currently the most important traffic contract parameter [13]. To guarantee the same CLRs to both cell streams, the background traffic is replicated and sent simultaneously to both multiplexing stages. The multimedia workstations are connected as two point-to-point communications. One of the connections uses AAL5 to transmit the MPEG-2 bit stream while the other uses the proposed AAL.

The traffic under test (TUT) consists of a ski sequence of 1000 frames (720 × 576) encoded with MPEG-2 at 4 Mbit/s with a structure of 12 images per GOP with two B pictures between every reference picture and a single slice per line. The encoded bitstream is encapsulated into TS packets prior to be sent to the network. The resulting traffic is sent to the switch at a constant cell rate where it is multiplexed with the background traffic. Since the switch buffer is limited in size, some of the TUT or background cells may be lost. The TUT is then routed to the receiver end system where the data is reassembled prior to decoding. The background traffic is assumed not to interfere further with the TUT and thus is directly routed to a traffic sink after leaving the switching stage.

The transmission of video over AAL5 is based on the approved ATM Forum Video on Demand specification [3]. This document describes the encapsulation of MPEG-2 TS packets into AAL5-SDUs (Service Data Units). This scheme packetizes two single program transport streams (SPTS) packets regardless of their information contents, being of audio, video or timing nature into a single AAL5-SDU. The AAL5 adds its 8 byte trailer and the resulting AAL5-PDU is segmented into 8 ATM cells without any padding (see Fig. 2). This encapsulation method may support other PDU sizes, 376 being the default. However, some padding may be necessary to align larger AAL5-PDUs to ATM cell boundaries.

The transmission of video over the proposed AAL is based on the segmentation described in Sec. 3.1. Two TS packets are passed from the Network Adaptation layer to the AAL. The AAL-PDU is then segmented into 47-byte payloads giving exactly 8 ATM cells. When FEC is applied, we add a single redundancy cell obtained by XORing the 8 data cells. A single cell loss per PDU can be recovered.

4.2 A Perceptual Quality Measure as a Performance Criterion

Several studies have shown that a correct estimation of subjective quality has to incorporate some modeling of the Human Visual System [14]. A spatio-temporal model of human vision has been developed for the assessment of video coding quality [15, 16, 17]. The model is based on the following properties of human vision:

- The responses of the neurons in the primary visual cortex are band limited. The human visual system has a collection of mechanisms or detectors (termed channels) that mediate perception. A channel is characterized by a localization in spatial frequency, spatial orientation and temporal frequency. The responses of the channels are simulated by a three-dimensional filter bank.
- In a first approximation, the channels can be considered to be independent. Perception can thus be predicted channel by channel without interaction.
- Human sensitivity to contrast is a function of frequency and orientation. The *contrast sensitivity function* (CSF), quantizes this phenomenon, by specifying the detection threshold for a stimulus as a function of frequency.
- Visual masking accounts for inter-stimuli interferences. The presence of a background stimulus modifies the perception of a foreground stimulus: masking corresponds to a modification of the detection threshold of the foreground according to the local contrast of the background.

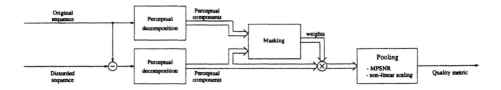

Fig. 8. *Moving Pictures Quality Metric (MPQM) block diagram*

The vision model described in [15] has been used to build a computational quality metric for moving pictures [16] which proved to behave consistently with human judgments. Basically, the metric, termed Moving Pictures Quality Metric (MPQM), first decomposes an original sequence and a distorted version of it into perceptual components. The quality measure is then computed, accounting for contrast sensitivity and masking (see Fig. 8) and scaled from 1 to 5 as described in Tbl. 1 [18].

Rating	Impairment	Quality
5	Imperceptible	Excellent
4	Perceptible, not annoying	Good
3	Slightly annoying	Fair
2	Annoying	Poor
1	Very annoying	Bad

Table 1. *Quality scale that is often used for subjective testing in the engineering community*

Recently, an improved metric, termed Normalized Video Fidelity Metric (NVFM), based on a finer modeling of vision, has been introduced in [19]. This new metric adds a modeling of the saturation characteristic of the cortical cells' responses and a modeling of inter-channel masking. It is an extension of a still-picture model developed by Teo&Heeger [20].

4.3 Performance Measurements and Analysis

To study the performance of the proposed AAL, we have carried out simulations based on the setup described in Sec. 4.1 for different background loads varying between 79 and 86%. The background sources generate a balanced load. Figure 9 shows three curves with the CLRs measured at the receiver, which for AAL5 includes all the data lost due to the packet discard mechanism.

The CLR measured for the new AAL is equal to the network loss ratio since there is no extra discard of information at the AAL. For all the experiments, we obtain an improvement factor close to 8 in terms of CLR. This is basically due to the size in cells of the PDUs transmitted and suggests a low correlation in the cell loss process. It is interesting to note that using larger PDU sizes with the proposed AAL will not affect the CLR measured. This will not be the case if AAL5 is used since the impact of cell losses will be amplified by the packet discard mechanism albeit the probability of having more than one cell lost within the same packet will also increase. To perform the selective FEC experiment, we have tuned the network adaptation layer to protect three types of MPEG-2 headers, namely: sequence, picture and Packet Elementary Stream (PES) headers. We have protected a PDU with a single cell as described in Sec. 4.1. The overhead due to this mechanism is very small (4.75×10^{-3}) since there is little amount of loss sensitive data. Consequently, the results do not show

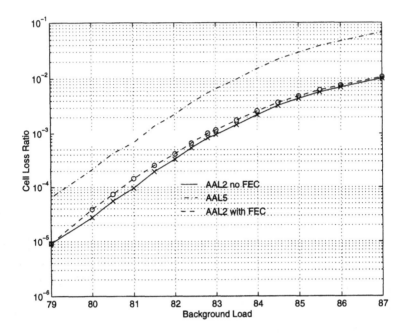

Fig. 9. *Cell Loss Ratios seen by receivers*

any particular improvement on the CLRs. They are even a little bit higher than in the no FEC case. However, the measurements depicted in Tbl. 2 show that a certain number of cells and therefore headers have been recovered. This means that syntactic information that may have been lost in the other two experiments will be available for decoding.

Load (%)	79	80	81	82	83	84	85	86	87
Recovered Cells	0	2	7	15	46	97	147	244	339

Table 2. *Recovery Efficiency of the Selective FEC Mechanims*

One of the major difficulties with video applications is the mapping of data loss impact into the quality perceived by the user. Due to the different types of data carried in a multimedia stream, the impact of loss may be different. To assess the efficiency of the proposed mechanisms, we have used a perceptual quality metric described in Sec. 4.2. We have also applied error concealment techniques [10] that reduce the impact of cell losses on the perception of decoded video sequences. Using these techniques allows us, for a given confidence interval, to estimate the mean value of the perceptual quality on a lower number of frames.

The quality estimation has been performed on 100 frames.

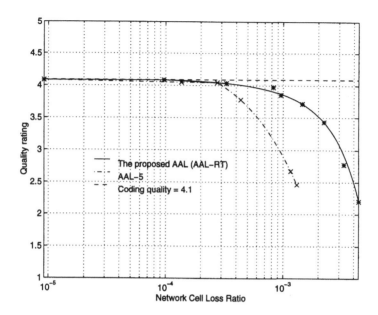

Fig. 10. *Quality Rating vs. Network CLR for AAL5 and New AAL*

Figure 10 shows the MPQM quality assessment as a function of the cell loss ratio for the transmission of MPEG-2 coded streams over AAL5 and the proposed AAL. While increasing the CLR, the perceptual quality obtained with the proposed AAL remains nearly constant at the maximum value corresponding to the quality of the MPEG-2 coded sequence compared to the original not coded one (horizontal dotted line). This is due to relatively sparse cell losses easily masked by the use of error concealment techniques which therefore leads to a very low impact onto the quality. Beyond a CLR close to 3×10^{-3}, the perceptual quality drops smoothly since only a few frames have been lost, keeping efficient the error masking. Conversely, the curve obtained with AAL5 shows a different behavior. Two regions separated by a critical CLR can be distinguished: a first one with almost no quality degradation and a second one where the quality drops fast. This is mainly due to the packet discard mechanism. Considering the fact that for a single cell loss two TS packets are discarded, it is clear that the probability of losing syntactic information is higher than for the proposed AAL. Therefore more frames and PES may be lost. Also, if the data has been lost within the picture, the *holes* will be bigger and the masking less efficient than for the proposed AAL. Having less information available, the concealment mechanisms are less efficient leading to a faster quality degradation. Indeed, beyond the critical CLR, for a given loss ratio we achieve a significant gain in terms of perceived quality.

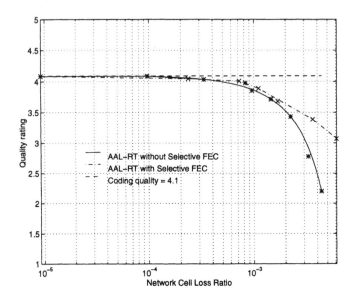

Fig. 11. *Quality Rating vs. Network CLR for New AAL and New AAL with Selective FEC*

Figure 11 presents the gain in perceptual quality we have obtained by protecting syntactic information, namely, sequence, frame and PES headers with the selective FEC mechanism. The utilization of selective data protection that generates very small overhead (0.47%) reduces the slope of the curve, hence, smoothing out even more the quality degradation. Under severe cell loss conditions, the transmission with selective FEC proves to be very robust. It is worth to note that we have applied a very simple protection scheme that proves to be efficient. Using more elaborated data protection algorithms may improve even more the quality degradation figures while still having a small FEC overhead.

5 Conclusions

We have presented in this paper mechanisms at the AAL to improve the transmission of real-time multimedia applications. We increase the cell error detection resolution to the cell level. We also propose a cell-based selective FEC mechanism to recover from cell losses. A basic network adaptation layer has also been presented that covers the segmentation of packets into fixed size PDUs and a selective FEC_request mechanism. Although network adaptation layer mechanisms are generic they necessarily have to be specific to the application since the request of FEC data relies on the knowledge of the information to be transmitted. Given the difficulty to map cell losses onto image quality, we have used a perceptual quality metric as a tool to evaluate network performance. We have carried out experiments with MPEG-2 video streams. The results obtained show an improvement of the cell loss figures for the simulations with and without selective FEC compared to current transmission methods as defined by the ATM Forum. From a perceptual point of view, the proposed AAL achieves a graceful quality degradation compared to AAL5 which shows a critical CLR value beyond which quality drops very fast. The application of a low overhead selective FEC scheme does not show any significant CLR improvement however it smooths out even more the quality degradation observed under severe CLR ratios. The FEC mechanism applied which protects a subset of the syntactic information, even though it is very basic, gives good results. Future experiments will be done with improved data protection schemes [21] to increase the robustness of audiovisual applications to data loss.

References

1. Thomas Stock and Reto Gruenenfelder, "Frame Loss vs. Cell Loss in ATM Concentrators and Multiplexing Units", in *EFOC&N'94*, 1994.
2. Israel Cidon, Roch Guerin, Asad Khamisy, and Kumar N. Sivarajan, "Cell versus Message Level Performances in ATM Networks", in *First International ATM Traffic Expert Symposium*, Basel, Switzerland, Apr. 1995.
3. The ATM Forum, "Video on Demand Specification 1.0", Jan. 1996.
4. Xavier Garcia Adanez, Andrea Basso, and Jean-Pierre Hubaux, "Study of AAL5 and a New AAL Segmentation Mechanism for MPEG-2 Video over ATM", in *7th Workshop in Packet Video*, Brisbane, Australia, Mar. 1996, available on http://tcomwww.epfl.ch/~garcia.
5. A.J. McAuley, "Reliable Broadband Communications Using a Burst Erasure Correcting Code", in *ACM SIGCOMM'90*, pp. 297–306, Philadelphia, PA, 1990. ACM, Academic Press.
6. ISO/IEC JTC-1, editor, *Information Technology - Generic Coding of Moving Pictures and Associated Audio Information - Part 1: Systems Specification*, ISO/IEC JTC-1, 1994.
7. Olivier Verscheure and Xavier Garcia Adanez, "Perceptual Quality Metric as a Performance Tool for ATM Adaptation of MPEG-2 based Multimedia Applications", in *EUNICE Summer School on Telecommunications Services*, Lausanne, Switzerland, September 1996, available on http://tcomwww.epfl.ch/~garcia.
8. H. Leopold, G. Blair, A. Campbell, G. Coulson, P. Dark, F. Garcia, D. Hutchinson, N. Singer, and N. Williams, "Distributed Multimedia Communication System Requirements", OSI 95 Consortium, May 1992.
9. ITU-T, editor, *Recommendation I.363 B-ISDN ATM Adaptation Layer (AAL) Specification*, ITU-T, Geneva, Mar. 1993.
10. Carlos Lopez Fernandez, Andrea Basso, and Jean-Pierre Hubaux, "Error Concealment and Early Resynchronization Techniques for MPEG-2 Video Streams Damaged by Transmission over ATM Networks", in *SPIE*, Brisbane, Mar. 1996.
11. Mohammad Ghanbari and Vassilis Seferidis, "Cell Loss Concealment in ATM Video Codecs", *IEEE Transactions on Circuits and Systems for Video Technology*, vol. 3, June 1993.
12. ISO/IEC JTC-1, editor, *Information Technology - Generic Coding of Moving Pictures and Associated Audio Information - Part 2: Video*, ISO/IEC JTC-1, 1994.

13. Rastislav Slosiar, *Performance Analysis Methods of ATM-Based Broadband Access Networks Using Stochastic Traffic Models*, PhD thesis, TCOM Laboratory, Swiss Federal Institute of Technology, CH 1015 Lausanne, Switzerland, 1995.

14. Serge Comes, *Les traitements perceptifs d'images numerisées*, PhD thesis, Université Catholique de Louvain, 1995.

15. Christian J. van den Branden Lambrecht, "A Working Spatio-Temporal Model of the Human Visual System for Image Restoration and Quality Assessment Applications", *in ICASSP*, pp. 2293–2296, Atlanta, GA, May 7-10 1996.

16. Christian J. van den Branden Lambrecht and Olivier Verscheure, "Perceptual Quality Measure using a Spatio-Temporal Model of the Human Visual System", *in Proceedings of the SPIE*, vol. 2668, pp. 450–461, San Jose, CA, January 28 - February 2 1996, available on http://tcomwww.epfl.ch/~verscheu.

17. Christian J. van den Branden Lambrecht, *Perceptual Models and Architectures for Video Coding Applications*, PhD thesis, Swiss Federal Institute of Technology, CH 1015 Lausanne, Switzerland, 1996, available on http://ltswww.epfl.ch/pub_files/vdb/.

18. M. Ardito, M. Barbero, M. Stroppiana, and M. Visca, "Compression and Quality", *in Proceedings of the International Workshop on HDTV 94*, Brisbane, Australia, October 1994.

19. Pär Lindh and Christian J. van den Branden Lambrecht, "Efficient Spatio-Temporal Decomposition for Perceptual Processing of Video Sequences", *in ICIP*, Lausanne, Switzerland, September 16-19 1996, available on http://ltswww.epfl.ch/~vdb.

20. David J. Heeger and Patrick C. Teo, "A Model of Perceptual Image Fidelity", *in ICIP*, pp. 343–345, Washington, DC, October 23-26 1995.

21. Olivier Verscheure, "Perceptual Video Quality and Activity Metrics : Impact on MPEG-2 based Video Services Optimization", *in COST-237 workshop*, Barcelona, Spain, November 1996, available on http://tcomwww.epfl.ch/~verscheu.

Multimedia Applications on a Unix SVR4 Kernel: Performance Study

Daniela Bourges Waldegg *   Naceur Lagha   Jean-Pierre Le Narzul

Telecom Bretagne
Networks and Multimedia Services Department
BP 78, 35512 Cesson-Sévigné
France
E-mail: {bourges, lenarzul, lagha}@rennes.enst-bretagne.fr

Abstract. End-system resource management is a key issue for obtaining satisfying QoS in multimedia applications. As these applications are being incorporated into workstations, problems regarding efficient and well suited operating system mechanisms to cope with multimedia data constraints have been outlined. In this paper we present a runtime kernel analysis of an adaptive multimedia application dealing with compressed audio and video, which is structured as different communicating processes, running on Solaris 2.4. These processes are not independent, and their interactions determine the application's overall performance. We show that even if application structure is optimised, the fact that synchronisation between processes exists but is not considered by the process scheduling policy, results in unnecessary data copying and image rejection and consequently poor performances. Our measurements of application performance highlight the need for a system level synchronisation support.

1 Introduction

Multimedia services need end-to-end quality of service (QoS) support. This means that adapted resource management mechanisms have to be offered by networks, but also by end-systems. At the end-system, the operating system is responsible for the allocation of local resources.

Multimedia applications are increasingly being incorporated into classical Unix operating systems. However, their specific characteristics have to be taken into account by the OS in order to obtain satisfying performances. Multimedia applications deal with different types of inter-related data, particularly continuous media like audio and video. Compared to traditional data, continuous media processing imposes new constraints to the operating system. In addition, multimedia applications co-exist with traditional Unix processes, so an adapted resource management scheme satisfying both kinds of applications is necessary.

Video streams represent the more constraining data, because of their volume and their required image presentation rate. Video decompression can be accomplished by hardware components or in software. Software video decompression is very demanding in terms of CPU time, but it is a flexible solution for versatile environments. Con-

---

* Supported by a grant from the Mexican National Council for Science and Technology (CONACYT), under the SFERE/CONACYT Cooperation program

tinuous media also present stringent timing relationships, inside each stream (intra-stream synchronisation), but also between different streams (inter-stream synchronisation), as is the case for video and associated audio [9].

In this paper, we analyse the behaviour of the Solaris 2.4 kernel during execution of a public domain multimedia application which deals with synchronized and compressed audio and video. The purpose of this analysis is to determine the impact of OS mechanisms on this multimedia application's performance, and to distinguish between the effects of application structure and the problems caused by the kernel itself.

We focus on the Solaris 2.4 operating system, which is the Unix SVR4 from Sun, for several reasons. SVR4 systems are widely used in workstations, and the Solaris 2.4 implementation introduces important structural features that are promising for multimedia applications.

The need for multimedia application support in operating systems has been treated in several research projects [3] [4] [6] [10]. Real-time characteristics of multimedia data have been recognized, regarding the presentation of a single flow (rather than the presentation of a set of inter-synchronized flows). The main challenge of multimedia OS support is to find a general resource management scheme that satisfies different applications' QoS requirements. Scheduling policies are a major problem, because a trade-off has to be found between fairness, soft real-time processing and fast response for human interaction. The need for real-time scheduling to support multimedia has been widely pointed out, but the influence of inter-media synchronisation in scheduling policies is not yet well studied.

Other problems regarding OS support for multimedia include data copying, which has to be minimised due to the large volumes of information inherent to multimedia. System overhead, like costly context switches, needs also to be reduced in order to satisfy time requirements and to take a better advantage of hardware capacities. Protocol processing, in addition, cannot be independent of the application state and needs also to avoid unnecessary data copying.

Some interesting new abstractions and architectural principles to solve these problems have been proposed in [2]. Coulson *et al* have worked with the Chorus microkernel to define a distributed multimedia platform supporting ATM communications; in their architectural propositions, they include an extension to the split-level scheduling scheme proposed by Govindan *et al* [6], which is based on user-level schedulers for threads, kernel level schedulers for virtual address spaces and an asynchronous communication of threads' states between them.

It is also interesting to observe how existing operating systems behave on execution of multimedia applications, to determine if multimedia adaptations of such systems are possible. The Solaris OS has introduced interesting features in its architecture that can be very useful. Its kernel is completely multithreaded [5], and user-level threads are also supported. Several scheduling classes are defined [8], particularly a real-time class. But even with these features, there is still a large gap to achieve good application performance. For instance, problems like inter-media synchronisation are not well managed by the fixed priority real-time scheduling policy. Our goal is to analyse the behaviour of the Solaris 2.4 kernel on executing an application with severe constraints, particularly inter-media synchronisation.

In this context of application performance observation, Nieh *et al* [11] analysed the Solaris 2.2 kernel, concluding that it was not adapted for running multimedia applications, and proposed a modification to the time sharing scheduling class implementation that was incorporated into Solaris 2.3. This study treated the execution of different kinds of independent processes (an interactive process, a CPU intensive background process and a video displaying process) rather than a multimedia application as a whole.

In our study, we focus on CPU behaviour problems that arise when compressed and synchronized data have to be managed, in a communicating process structure. Particular interest is given to the fact that processes are scheduled independently and are at the same time subject to communication and synchronisation interactions that determine the overall application performance.

The rest of the paper is organized as follows: in section 2, we will introduce the application we used in our study: the MPEG-1 system decoder/player from Boston University. Then, in section 3, we will describe our experimental testbed and the observed parameters. Finally, we will comment the experimental results, and we will discuss future trends.

2 Overview of the MCL MPEG-1 system decoder/player

The application chosen for this study is the MPEG-1 system decoder/player developed at the Boston University Multimedia Communications Laboratory (MCL) [1]. This application decodes and plays a compressed video stream and associated audio, adapting the video frame rate to runtime conditions such as system load and synchronisation with the audio stream.

The application is interesting for several reasons. It uses the MPEG standard, which is the major standard for compressed audio and video in multimedia applications, so it gives an outlook on the software MPEG decoding behaviour. It is an all-software implementation, which is very demanding in terms of CPU time, and synchronisation relationships between audio and video need to be respected. These two characteristics lead to a kernel analysis under very constraining situations (typical of multimedia applications). And last, the application was built using a "building blocks" approach, as it used two existing applications (the Berkeley Mpeg-play2 video decoder and the Maplay2 audio decoder developed at the Berlin University of Technology); this way of assembling an application makes software re-utilisation possible, and allows programmers to easily integrate different decoders and to follow the evolution of one specific decoding algorithm. However, it raises several architectural problems, as will be further seen.

2.1 MPEG

MPEG is a video and audio compression standard specified by a joint ISO and IEC international committee (the *Motion Picture Experts Group*). MPEG-1, the first draft proposal for this standard, was produced in 1990 and reached the status of International Standard in 1992 [7]. This first standard was defined for the storage and retrieval of movies at up to 1.5 Mbits/s. MPEG-2, a second version of the standard with a wider range of applicability, was specified in 1994. We will further restrain our description to MPEG-1, since it is this version which is implemented by the MCL decoder/player.

The MPEG-1 standard comprises three parts:

- MPEG-video specifies the coded bitstream format and the compression techniques for digital video. Video compression reduces spatial redundancy by means of a Discrete Cosine Transform technique and a weighted adaptive quantization. Temporal redundancy is exploited also, using a motion compensation technique for inter-frame coding. Three types of frames (images) are defined: I (Intra-coded) images are coded without reference to other images; P (Predictive) images are coded with reference to past I or P images, and B (Bidirectional) images make reference to a past image and a future image of type I or P. Images are organized as GoPs (Groups of Pictures), starting with an I image, followed by an arrangement of P and B images (IBBPBBI, for instance). The best compression rate is achieved for B images; the GoP structure is then a trade-off between compression rate (more B images) and robustness (I images stop error propagations).

- MPEG-audio specifies the bitstream structure and compression technique for audio data. Sampling is defined at 32, 44.1 or 48 KHz, and samples are linear encoded using a 16 bits PCM. Two audio channels are defined. Audio compression is based on the elimination of low amplitude components that are masked to the human ear because of their proximity in time or frequency to higher amplitude components.

- MPEG-system defines a structure for multiplexing up to 16 MPEG-video streams and 32 MPEG-audio streams, and the corresponding synchronisation information for decoding and playback. The structure is called a *Program Stream*; it is a combination of Packetized Elementary Streams (PES) sharing a common time base. The Program Stream is structured in packs which contain a variable number of variable length PES packets. The header of each pack contains the common clock information, whereas per stream presentation and decoding time stamps are contained in PES headers (decoding time may precede presentation time because of the inter-frame dependency).

2.2 Application Structure

The runtime structure of the MCL MPEG-1 system decoder/player (*System-play* from now on) is shown in Figure 1. It consists of several communicating processes: one for each video stream, one for each audio stream, a demultiplexer and an audio mixer. In theory, the application can decode up to 16 video streams (16 video processes) and 32 audio streams. For practical reasons (the capacity limitations of a workstation), our study will only consider the case of a single video stream and its corresponding audio stream.

In the original application, processes communicate data streams through standard Unix pipes. In order to minimize communication overhead, we implemented shared memory FIFOs (called *data shared memory*) instead of pipes. Synchronisation information from the MPEG system stream is stored in a semaphore controlled shared

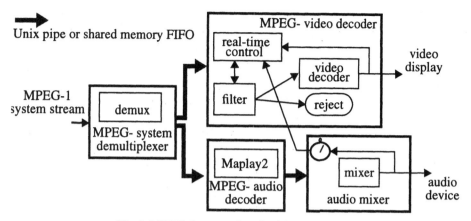

Fig. 1. MPEG-1 system decoder/player structure

memory segment (called *synchronisation shared memory*), and is used to synchronize video with respect to audio.

In our study, we analysed both versions of System-play. The basic application principles are the same for both modes of process communication, so in the remainder of this section we will call pipes and data shared memories *communication channels* to ease our explanation.

When all processes are initialized, the application works as follows:

• the demultiplexer reads the MPEG-system stream from the disk, it identifies the different sequence codes and writes packets to the appropriate decoder's channel. That is, when it reads a packet from the video stream, it writes it to the video decoder communication channel, and similarly for audio stream packets. Time stamps extracted from the packet headers are written into the synchronisation shared memory,

• the video process executes the Berkeley Mpeg-play2 video decoder, which rejects out of synchronisation frames (adaptive playback). This feature is based upon the periodic execution of an algorithm that determines a *playing phase* (explained below). Video packets are read from the demultiplexer communication channel; when a picture start code is detected, a decision is taken based on the current playing phase and the frame type; if the frame is accepted, it is decoded and written into the X server frame buffer for display. If it is rejected, its packets are read and thrown away until the next picture start code arrives,

• the audio process executes the Maplay2 audio decoder. It reads from the demultiplexer communication channel and writes decoded audio packets into the audio mixer communication channel,

• the audio mixer has a double function: it reads the different audio streams from audio decoder communication channels and mixes them (which doesn't apply in our case with only one audio stream) and writes into the audio device buffer. The second function is to update the audio clock (by the number of written audio packets) in the synchronisation shared memory,

• every second, a software interrupt occurs in order to update the playing phase. This interrupt is handled by the video process itself (the *real-time control* block). The playing phase refers to the state of the playing video stream with respect to the audio stream, since the latter acts as the master clock. The playing state is computed by a comparison between the audio clock and a video clock (incremented each time a frame is received by the video decoder, independently if it is later rejected or decoded); it determines the type and number of frames that are to be played. The algorithm takes into account MPEG-video frame dependencies, so it first rejects B frames (since no frame depends upon a B frame, only the B frame will be lost), then P frames (which implies B frames are lost) and I frames at last.

The replacement of pipes by shared memory FIFOs is aimed at eliminating physical copies between kernel address space and user address space in read operations, since large volumes of data have to be transmitted (particularly video data). So, with a shared memory approach, the video and audio processes will directly read in the memory segment where the demultiplexer placed data (in fact, a read operation only consists of getting the correct buffer pointer). Similarly, the mixer reads data from the FIFO in which the audio process has written it.

3 Experimental Design

The aim of our study is to observe the Solaris 2.4 kernel behaviour on execution of System-play, and its effects on the application's performance. To do so, we defined a set of parameters to measure based on both application quality and kernel execution. We carried out the same tests for both versions of the application. Our goal is not to compare the two communication mechanisms, but to distinguish when problems come from the application structure and from the OS kernel, mainly from scheduling mechanisms.

The tests consisted of controlled executions of the applications over a Solaris 2.4 kernel capable of logging kernel event information into a tracing buffer.

3.1 Parameters

Application performance is determined by the quality of video presentation and of audio presentation. Video quality can be described by displayed image rate and its variability. Audio quality depends on the continuity of audio presentation.

Video processing is more demanding of CPU cycles, since it has to perform expensive calculations on important data volumes. So video quality is the most delicate part of application performance. Audio quality is generally good, because audio decoding is not very costly, and if the audio device buffer doesn't underflow, periodical playback of audio samples is guaranteed by the audio device. Therefore, application quality is mainly determined by the displayed image rate.

To observe the relationship between kernel behaviour and application performance, we defined the following parameters:

• decoding time for displayed frames: the elapsed real time between the reading of a picture start code and its writing into the X server frame buffer,

• effective decoding time for displayed frames: the effective CPU time used to decode a frame, that is, without considering the inter-task time where the video process is not holding the CPU; this parameter reflects in fact the CPU capacity (since it does not consider OS overhead).

Finally, we defined a set of parameters that represent possible causes of application performance degradation due to the kernel internals:

• context switches per displayed frame: the number of context switches that occur between the reading of a picture start code and its writing into the frame buffer via the X Server. In other words, the number of schedulings the video process needs to decode a given frame,

• total CPU occupation time and task durations for each process: the time during which the process holds the CPU (we call *task* the interval in which a process is in the running state, holding the CPU)

• inter-task time for each process: the time during which the process waits on a ready-to-run queue, preempted by a higher priority process, or in a sleep queue.

• blocking times for the video process: the time during which the video process is asleep waiting for a signal or blocked by an input/output.

3.2 Tracing mechanism

To log kernel activity, we enabled the Solaris 2.4 kernel tracing mechanism, which is based on tracepoints inserted in the kernel code. During runtime, when a tracepoint is encountered, the running kernel thread writes a trace record in a trace buffer; this trace record contains information about the encountered event, as well as a time stamp relative to the last tracepoint. The time stamp has a microsecond precision.

The tracing mechanism is controlled from user address space by means of a system call; a user process can reset, enable and disable tracing of desired events, choose a set of processes to trace, resume and pause tracing and flush the contents of the kernel trace buffer into a file.

The tracing mechanism produces a raw trace file; a number of tools were developed in order to decode the binary file, and filter the events depending on the observed parameters.

3.3 Experimental Conditions

Tests were performed on a Sparc Station 20 with a single 50 MHz processor and a 32 Mbytes primary memory, running Solaris 2.4. Both versions of System-play were repeatedly executed. They were slightly modified in order to control the tracing mechanism and to log information about video frame decoding.

User processes running during the tests were limited to the four System-play processes, the XWindow server and the console. The System-play user controls were not used to prevent the mouse motion from preempting other processes.

Solaris 2.4 supports several scheduling classes [8], each one with a different scheduling policy. The scheduler bases its decisions on a global priority vector, in which the different classes are mapped. The real-time class (mapped at the highest global priorities) uses a fixed priority policy. The time-sharing class (mapped at the lowest global priorities), uses a dynamic priority policy, with bigger time quanta for lower priority processes; a time-sharing process priority will depend on its CPU utilisation (for higher utilisation, lower priority).

All user processes in the experiments were scheduled in the time sharing class; the usage of the real time class was studied, but the fact that System-play processes are inter-dependent leads to a system-wide deadlock (because a ready real-time class process prevents any other system or time-sharing process to run, and the video process is almost always ready to run); a similar situation was reported by [11].

We used an MPEG-system sequence containing 2003 SIF images (352X288 pixels) at a rate of 25 images per second (80.12 seconds of playback). The sequence bitrate is 428 kbits/s, where 391 kbits/s correspond to the video stream and 33 kbits/s to the audio stream. The sequence contains an interview of the MCL team.

4 Results

In the following, we will discuss the measuring results for typical executions of both versions of System-play, that is, pipes and shared memory. The pipes version executions are more unstable than those from the shared memory version. The measures we show for it correspond to executions where no significant problems occurred (like large periods with no displayed images).

4.1 Video Quality and Kernel Events

Table 1 summarizes the video quality parameters and related kernel measurements. The image animation enhancement obtained by shared memory communications is significant: 10 images/s compared to 1.4 images/s for the pipes version. The instantaneous displayed image rate is plotted in Figure 2. As the plots show, in both cases the image rate is very variable, which results in a changing speed of video presentation.

Table 1. Video quality parameters

Application	total displayed images	image rate (images/s)	decoding time (ms)		effective decoding time(ms)		context switches per displayed frame
		average	average	standard deviation	average	standard deviation	average
System-play pipes	112	1.4	308.10	102.04	252.33	67.29	6.10
System-play shared memory	816	10.1	62.12	51.75	48.08	18.33	1.08

A parameter that relates video quality and kernel activity is the time used to decode displayed frames. These measurements are shown in Figure 3. For a video stream to display 25 images per second, the decoding time for each image has to be less than 40

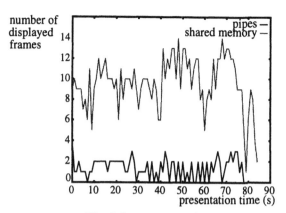

number of
displayed
frames

Fig. 2. Instantaneous image rate

ms. For the pipes version (Figure 3.a), decoding times are an order of magnitude higher than this ideal time; this causes frame rejection not only because of the CPU not being able to decode them in time, but also because of frame dependency. The effective time is also greater than the ideal 40 ms; this means that even if the video process could run alone, it would never display all its images (note that the effective time includes data communication, and several read operations are necessary to load the complete frame).

The same parameters for the shared memory version are plotted in Figure 3.b. Decoding times are still over 40 ms, but they fall in the same order of magnitude. Effective decoding times are significantly smaller than for the pipes version, going sometimes under the 40 ms limit. This means that in some occasions, the CPU is in fact capable of decoding frames in time. So even if data communication's effects are minimized using shared memory, system overhead and scheduling delays still cause a considerable performance loss. If the system takes longer to decode a given frame, subsequent frames will be rejected; rejecting many frames in a row causes an undesired blocking effect that leads to a greater frame rejection. So even if frames have to be rejected, the dispersion in time of these rejections is meaningful for final performance.

Another interesting parameter is the number of context switches the video process is subject to during frame decoding. This parameter is not necessarily related to the time the process waits to be rescheduled. For the pipes version (Figure 3.c), there is, in average, six context switches between the reading of the image start code and its display, with peaks of up to 18. In the shared memory version, the average context switching has been reduced to one (Figure 3.d), even if peaks of up to 15 are still present (the persistence of peaks is due to audio processing periods). There are more average context switches in the pipes version because of the pipe's limited size, as the demultiplexer process preempts the video process more often in order to write data, but it sleeps as soon as the pipe is full.

One of the reasons for the video presentation enhancement with the shared memory is the elimination of the data copying overhead, which lets CPU time be used in com-

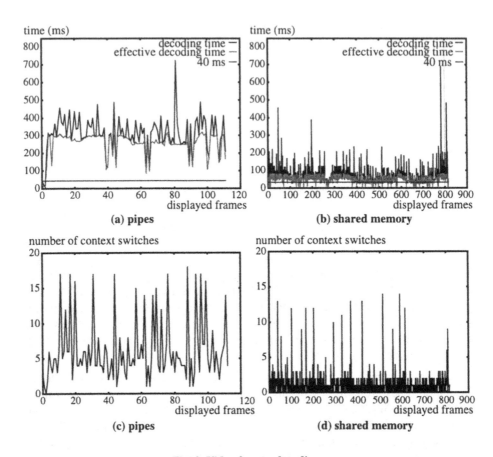

Fig. 3. Video frames decoding

putations rather than communications. Pipes represent an important overhead caused by data copying and, in addition, they are more limited in size than shared memory segments, so more context switches are needed for the transfer of a given amount of data.

To conclude this subsection, we emphasize on the considerable overhead the video process is subject to in order to decode a frame (that is, the difference between decoding time and effective decoding time). Once the data copying overhead eliminated, the image rate is enhanced. But in this case, the Sparc 20 CPU is though capable of decoding more frames that it actually does (since effective decoding times are near 40 ms). But the system as well as the synchronisation mechanism introduce important and variable delays; a decoded frame will then be displayed lately leading to a rejection of the following frames.

4.2 CPU utilisation

Table 2 shows the CPU occupation rates for each System-play process. The amount of CPU assigned to the video process is naturally higher than the others. In the case of pipes, this amount is used mainly on data reading, while in the shared memory version, it serves to decode more frames. Although the amounts of CPU cycles assigned to each process are similar in both versions, the final application performance is very different.

It is to be noted that the time that is not used by the System-play processes, is used either by the XWindow server, or corresponds to an idle CPU due to undesired blocking effects (this will be explained in section 4.3).

Table 2. CPU utilisation

Application	CPU utilisation
System-play pipes total	75.98 %
video	61.99 %
audio	9.24 %
demux	2.37 %
mix	2.38 %
System-play shared memory total	82.45 %
video	70.25 %
audio	7.72 %
demux	1.43 %
mix	3.05 %

Table 3 shows the context switches for each process, and their values for task durations.

Table 3. Context Switches and Tasks

Application	number of context switches	task duration (ms)	
		average	standard deviation
System-play pipes			
video	1575	31.24	38.87
audio	1053	6.97	11.42
demux	904	2.06	1.19
mix	937	2.02	3.79
System-play shared memory			
video	2012	29.88	22.87
audio	642	10.29	11.76
demux	651	1.86	1.20
mix	221	11.82	23.57

According to the Solaris scheduling policy, process priorities change depending on their CPU utilisation, so the video process will rapidly reach the lowest priority. The highest priority System-play process is the demultiplexer, followed by the mixer and the audio process.

The video process is scheduled more times in the shared memory version than in the pipes version, but the average task time is higher in the pipes version.The average time the video process waits to be rescheduled is greater in the pipes version, that is, it is less frequently scheduled; this time is also very changing, as suggested by its standard deviation value. This has an impact on the frame reject algorithm. In this algorithm, the decision of playing a frame is taken at the reading of the picture start code, without taking into account the future decoding time which is very variable; if the video process is less frequently scheduled, by the time the frame is played, the video stream is behind of synchronisation so more frames are subsequently rejected.

The audio process, on the contrary, is less frequently scheduled in the shared memory version, and its average task duration is higher. This is also due to the overhead elimination achieved with the shared memory; the audio process in the pipes version is in fact scheduled more times only to perform data copies. The reduction of schedulings for the audio process allows a more frequent scheduling for the video process. The difference in number of context switches is also important for the mixer process. It is less scheduled in the shared memory version, but for a longer time. In fact, the behaviour of the mixer depends directly on the audio process, since it only waits for data to write it into the audio device.

Figure 4 shows the distributions of the CPU utilisation. The curves represent the probability for a process to be scheduled for a certain amount of time. The probability for the video process (Figure 4.a) to be scheduled for a small period is very high, whereas the probability to stay longer in the CPU decreases exponentially. This is due to the scheduling policy that privileges all other processes. Even though the video process gets the biggest portion of CPU time, the number of decoded frames depends also on the *instants* in which the process gets the CPU (because of its synchronisation with the audio stream). The video process distribution for the shared memory version is more linear, which explains the performance enhancement.

The audio process distribution (Figure 4.b) for the shared memory version reflects the fact that the process holds the CPU for a longer period each time it is scheduled, and that it is less scheduled for small periods. In the pipes case, the audio process is scheduled for longer periods at the end of the presentation (represented by the peak around 40 ms in the distribution plot), due to an audio device blocking.

The mixer's behaviour is similar to the audio process behaviour. Its distribution (Figure 4.d), shows also longer task periods in the shared memory version.

The demultiplexer (Figure 4.c) is basically the same in both versions; it is the System-play process with the highest priority, and the data copy between the process buffer in which the stream is stored and the pipe/shared memory was not eliminated. The only difference is that task periods are more regularly distributed in the shared memory version.

In Figure 5, the task and inter-task periods for the video process are plotted, in order to observe the progression of process scheduling during the total presentation time. Inter-task times include the intervals during which the process is blocked, or it has been preempted by a higher priority process and is waiting in a run queue. For the pipes version, scheduling frequency is very changing. There is an interval (between the

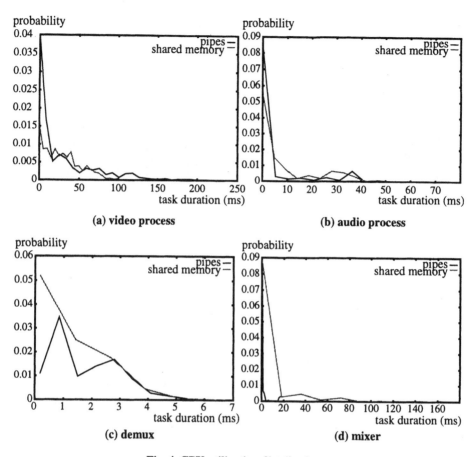

Fig. 4. CPU utilization distributions

30th and the 60th second of presentation) in which the process waits for long periods to be rescheduled. Although the total execution time of the video process is high, its irregular execution is one of the causes of poor performances.

To summarize, the results shown in this subsection explain the gain in performance from one version to another, which is a direct result of the more regular distribution of occupation times. This regular distribution of CPU cycles all through the presentation time allows the video process to decode more frames and leads to less bursty frame rejections. The video process more linear occupation time distribution in the shared memory version is due, on the one hand, to the reduction in video data copying overhead, and on the other, to better distributions of the other processes. This kind of occupation times regular distribution is though a desirable behaviour of the scheduling policy. Note however that this behaviour is not granted if the system executes processes other than the System-play ones.

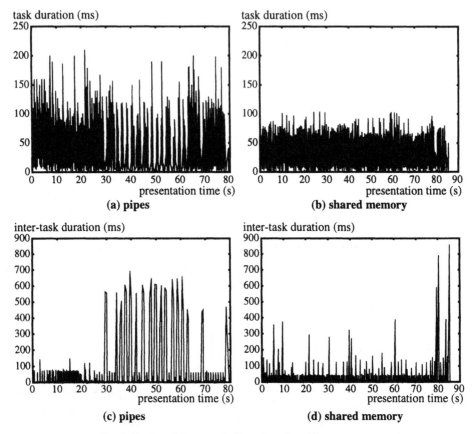

task duration (ms)

(a) pipes

task duration (ms)

(b) shared memory

inter-task duration (ms)

(c) pipes

inter-task duration (ms)

(d) shared memory

Fig. 5. Task and Inter-task durations for video process

4.3 Video Process Blocking

Process blockings can be produced by a blocking read or write operation or an explicit pause or sleep decided by the process.

The video process blocking results are shown in Table 4. The video process should never be blocked on a read operation, since it consumes data more slowly than the demultiplexer produces it. However, we observed that in the pipes version, it blocks for a total of almost 11 seconds (9% of the total playback time). These blockings occur when a large number of frames are rejected in a row, and the video process finds no data to read since the demultiplexer is sending data to the audio process.

This unwanted effect changed with the utilisation of shared memory. The blocking time is still significant (6 seconds, 5% of the total time), but blockings are due to another reason: the video stream clock gets too much ahead from the audio clock, that is, too much frames have been seen before the periodic interrupt, so the video process pauses until reception of the alarm signal that causes the execution of the phase algo-

Table 4. video process blockings

Application	number of blockings	blocking duration (ms)	
		mean	standard deviation
System-play pipes video	45	240.36	259.17
System-play shared memory video	31	201.84	213.48

rithm. In this case scheduling of the process ahead of synchronisation caused the blockings.

Video process blocking in the case of pipes gets worst due to an overflow of the audio device which receives bursts of audio data; when this happens while the demultiplexer is sending audio data, and the video process has no data, all the application is blocked (since the audio process cannot write into the mixer process pipe, as the mixer is blocked on a writing operation).

In summary, blockings of the video process in any case are unwanted. When they occur, even if the CPU is available, the video process is not able to continue decoding frames and it then rejects more frames as a consequence of getting out of synchronisation. In the case of pipes, blockings are related to the irregular scheduling of the process, which leads to bursts of rejected frames and an underflow of the receiving pipe. In the shared memory case, they are due to non-controlled scheduling instants.

5 Discussion

We have presented the measurement results of operating system efficiency in the presence of multimedia processing, which is very demanding of CPU cycles, represents large volumes of data and is very time sensitive. Processes in our study are interdependent, so they have specific temporal and data exchange synchronisation needs (as data handled by one process need to be processed and played out according to the state of another process' data). Temporal data synchronisation makes it difficult to apply classical real-time scheduling policies in which processes are considered independent.

As temporal and communication synchronisation represent severe constraints for the application execution, and therefore contribute to the performance degradation, it is useless to propose new OS mechanisms if the former are not optimised. In other words, unadapted application structure constitutes the performance bottleneck. This is the reason why we optimised the application structure in order to clearly distinguish its OS needs. In our case, the optimisation consisted in replacing pipes by shared memory FIFOs. An important gain in application performance was obtained with this modification, as we have shown in section 4. We have explained the differences in kernel behaviour for both cases, giving thus an insight on the system event evolution during MPEG-system playback. It is to be noted that the observed behaviour of the application can change if the MPEG stream is read from the network, as delay and jitter would influence the presentation development.

Even if performance is enhanced, degradations still persist after the application's structure modification. These are due to the independence between process scheduling and inter-media synchronisation, i.e., media synchronisation is not considered by the scheduling policy to assign execution priorities (since synchronisation is implemented at the application level). For instance, the video process is often scheduled only to reject frames, or to fall asleep because it has come ahead of schedule, as mentioned in subsection 4.3. If the system knew that processes treat inter-dependent data, it could base its scheduling decisions in these relations. Furthermore, even with an optimised structure, application performance would become unstable in presence of a variable system load, because changes in the execution of one process affect all others, as a *best-effort* policy is used.

Concerning application structure, we have several remarks. The approach used to build the System-play application is probably the most practical and intuitive one; it prevents from re-writing complex code for video and audio decoding, and follows the MPEG-system layered structure. But as we have shown, this structure causes many problems at kernel level. Pipes are definitely not well suited for multimedia data communication; shared memory FIFOs are better, but their implementation is not straightforward, and problems related to the scheduling policy still affect overall performance. The System-play synchronisation implementation is, on the other hand, very unstable and depends upon too many external factors, like system load. In addition, there is an unnecessary data transmission (and consequent context switching) between the demultiplexer and the video process with the down-scaling approach; a great part of this communication could be prevented if the video process informed the demultiplexer of its phase evolution so that the latter does not send frames that will not be decoded. This highlights the need for a multimedia application programming framework.

6 Perspectives

A desirable OS behaviour to cope with the outlined problems would take into account dependencies between data in different processes, when scheduling decisions are taken. In this context, we are currently working on the definition of a scheduling model that uses synchronisation parameters of groups of processes to assign execution priorities. The parameters definition follows a generic synchronisation specification that is translated into scheduling priorities. Applications need to inform the system on its quality of service needs prior to execution; we include in these needs the synchronisation relations between processes or threads.

7 References

[1] Boucher, J.A., Yaar, Z., Rubin, E.J., Palmer, J.D. and Little, T.D.C.- Design and Performance of a Multi-stream MPEG-I System Layer Encoder/Player. *In: Proceedings of the IS&T/SPIE Symposium on Electronic Imaging Science and Technology.*- San Jose, California, USA, February 1995

[2] Coulson, G. and Blair, G.- Architectural Principles and Techniques for Distributed Multimedia Application Support in Operating Systems. *Operating Systems Review*, ACM Press. Vol. 29, Number 4, pp 17-24, October 1995

[3] Coulson, G. and Blair, G.S.- Micro-kernel Support for Continuous Media in Distributed Systems, *Computer Networks and ISDN Systems*. Vol. 26, pp 1323-1341, Special Issue on Multimedia, 1994

[4] Cranor, C.D. and Parulkar, G.M.- Design of Universal Continuous Media I/O. *In Proceedings of the 4th International Workshop on Network and Operating System Support for Digital Audio and Video.* - Durham, New Hampshire, April 1995

[5] Eikhot, J.R., Kleiman, S.R., Barton, S., Faulkner, R., Shivalingiah, A., Smith, M., Stein, D., Voll, J., Weeks, M. and Williams, D.- Beyond Multiprocessing - Multithreading the SunOS Kernel. *Sun Technical Bulletin*.- August 1992

[6] Govindan, R., and Anderson, D.- Scheduling and IPC Mechanisms for Continuous Media. *In: Proceedings of the Thirteenth ACM Symposium on Operating Systems Principles*, SIGOPS, Vol. 25, pp 68-80.- Pacific Grove, California, USA, 1991

[7] ISO CD 11172.- Coding of Moving Pictures and Associated Audio for Digital Storage Media at up to 1.5 Mbits/s. 1992

[8] Khanna, S., Sebrée, M., Zolnowsky, J.- Realtime Scheduling in SunOS 5.0. *In: Proceedings of Usenix 92*.- Winter 1992.

[9] Little, T.D.C., Ghafoor, A., Chen, C.Y.R. and Berra, P.B. – Multimedia Synchronization. *IEEE Data Engineering Bulletin*, Vol. 14, n° 3, pp. 26–35, September 1991

[10] Mercer, C.W., Savage, S. and Tokuda, H.- Processor Capacity Reserves for Multimedia Operating Systems. *In Proceedings of the IEEE International Conference on Multimedia Computing and Systems.* May 1994

[11] Nieh, J., Hanko, J.C., Northcutt, J.D. and Wall, G.A. – SVR4UNIX Scheduler Unacceptable for Multimedia Applications. *In : Proceedings of the 4th International Workshop on Network and Operating System Support for Digital Audio and Video*, ed. by Shepherd, D., Blair, G., Coulson ,G., Davies, N. and Garcia, F., pp. 41–53. – Lancaster, UK, November 1993.

Perceptual Video Quality and Activity Metrics: Optimization of Video Service Based on MPEG-2 Encoding

Olivier Verscheure and Jean-Pierre Hubaux

Telecommunication Services Group - TCOM Laboratory
Swiss Federal Institute of Technology
CH-1015 Lausanne, Switzerland
E-Mail : verscheure@tcom.epfl.ch
URL : http://tcomwww.epfl.ch/~verscheu

Abstract. In this paper, we first present a video quality metric based on human vision. We then propose a new perceptual video local activity metric. These two metrics are of great interest for the optimization of both the transmission and the coding of video sequences. Consequently, next we motivate the use of these metrics in order to increase the end-user perceptual quality of MPEG-2 based video services by showing results on three different application fields.

1 Introduction

Quality assessment of image sequences has recently become an important issue and receives a large amount of attention from the multimedia communications community. For the past several years, a large number of audiovisual services has been emerging (e.g. Video on Demand (VOD), Interactive Distance Learning (IDL), home shopping, etc). Today, ATM technology, efficient compression techniques and other developments in telecommunications make it possible to offer such services. These services should be optimized to be proposed at very attractive prices (the user expects an adequate video quality at the lowest possible cost). One of the major issues facing this optimization is the utilization of both perceptual video quality and activity metrics. Such metrics, which present the advantage of being coherent with human judgements, have recently begun to emerge [1, 2, 3].

This work motivates the use of both video quality and activity metrics based on psychophysics. Such metrics are of great interest for both transmission and coding optimizations of video services.

This paper is divided in four sections. Section 2 first describes a spatio-temporal model of the human visual system from which a perceptual video quality metric is then derived. Section 3 presents two perceptual local video activity metrics. After some insight on the transmission and coding of MPEG-2 sequences, the impact of these quality and activity metrics on the optimization of MPEG-2 based video services is the subject of section 4. Section 5 finally concludes the paper and provides an outlook of how this work will be continued.

2 A Perceptual Video Quality Metric

Several studies have shown that a correct estimate of subjective quality has to incorporate some modeling of the Human Visual System [4]. A spatio-temporal model of human vision has been developed for the assessment of video coding quality [5, 3, 6]. This model is based on the following properties of human vision:

- The responses of the neurons in the primary visual cortex are band limited. The human visual system has a collection of mechanisms or detectors (termed "channels") that mediate perception. A channel is characterized by a localization in spatial frequency, spatial orientation and temporal frequency. Therefore, the responses of the channels can be simulated by a three-dimensional filter bank.
- In a first approximation, the channels can be considered to be independent. Perception can thus be predicted channel by channel without interaction.
- Human sensitivity to contrast is a function of both frequency and orientation. The *contrast sensitivity function* (CSF) quantizes this phenomenon by specifying the detection threshold for a stimulus as a function of frequency.
- Visual masking accounts for inter-stimuli interferences. The presence of a background stimulus modifies the perception of a foreground stimulus. Masking corresponds to a modification of the detection threshold of the foreground according to the local contrast of the background (see section 3.2).

The vision model described in [5] has been used to build a computational quality metric for moving pictures [3] which proved to behave consistently with human judgements. Basically, the metric, termed Moving Pictures Quality Metric (MPQM), first decomposes an original sequence and a distorted version of it into perceptual components by a Gabor filter bank. Indeed, psychovisual tests have shown that the profile of the perceptual channels is very close to Gabor functions. A channel-based distortion measure is then computed accounting for contrast sensitivity and masking. Finally, the data is pooled over the channels to compute the quality rating (by means of the Masked Peak Signal to Noise Ratio, MPSNR) which is then scaled from 1 to 5 as described in Tbl. 1 [7] (see Fig. 1).

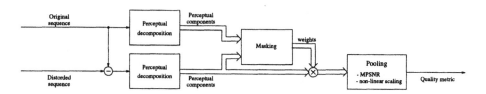

Fig. 1. *Moving Pictures Quality Metric (MPQM) block diagram.*

Recently, an improved metric, termed Normalized Video Fidelity Metric (NVFM), and based on a finer modeling of vision, has been introduced in [8].

Rating	Impairment	Quality
5	Imperceptible	Excellent
4	Perceptible, not annoying	Good
3	Slightly annoying	Fair
2	Annoying	Poor
1	Very annoying	Bad

Table 1. Quality scale that is often used for subjective testing in the engineering community

This new metric adds a modeling of the saturation characteristic of the cortical cells' responses and a modeling of inter-channel masking. It is an extension of a still-picture model developed by Teo & Heeger [9].

Figure 2 presents the MPQM quality assessment of MPEG-2 video for the Mobile & Calendar sequence as a function of the bit rate. This sequence has been encoded, with a software simulator of the Test Model v5 (TM-5) of MPEG-2 [10], as interlaced video with a constant Group of Pictures (GOP) of 12 frames with 2 B-pictures between every reference frame. An important result to be extracted from the graph is that the perceptual quality saturates at high bit rates. Increasing the bit rate may thus result, at some point, in a waste of bandwidth since the end user does not perceive an improvement in quality anymore.

In section 4, we will show results with the same behaviour on additional compressed video sequences.

3 Perceptual Local Video Activity Metrics

In the previous section, we have introduced a video quality metric based on a complete spatio-temporal model of the human visual system. Our objective is now to determine a way to classify sequence areas in terms of their relevance to human perception. In other words, we need to find a *local video activity measure* accounting for spatial and temporal perceptual activities that can be estimated independently of the transmission system and of the video encoding process; moreover, the proposed metric must not require any feedback mechanism.

This section proposes two new metrics, namely, the "Moving Pictures Activity Metric (MPAM)" and the "Perceptual Visibility Predictor (PVP)".

3.1 Moving Pictures Activity Metric (MPAM)

We propose this intuitive metric in order to introduce the perceptual video activity measure concept. We have considered the human visual system model presented in section 2 as a black box.

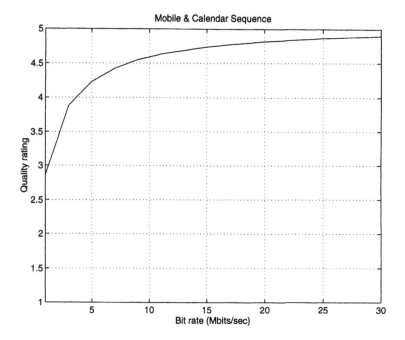

Fig. 2. *MPQM quality assessment for the Mobile & Calendar sequence as a function of the bit rate.*

Let us consider the block diagram presented in Fig. 3. A uniformly distributed white noise is first added to the original video sequence. For this purpose, a zero-mean uniform noise has been used taking values in $[-\alpha, +\alpha]$ with $1 \leq \alpha \leq 5$. We will see that the value chosen for α has only a little impact on the results. Both the sequence corrupted by the white noise and the original one are input to the vision model described in section 2. The output of the vision model is reconstructed to compute the perceived noise sequence which corresponds to the non-masked noise sequence. The procedure permits to account for pattern sensitivity, accounting for frequency sensitivity and visual masking. The final step consists in computing the non-perceived noise sequence, $npn(x, y, t)$, which is, by definition, the difference between the input noise and the perceived noise sequences.

In the $npn(x, y, t)$ [1] sequence, areas where the mean luminance value is high correspond to areas of the original sequence in which errors will most likely be masked by human vision. Likewise, areas where the mean luminance value is low correspond to zones in which visual masking does not occur.

---

[1] One can notice that, by definition, $npn(x, y, t)$ has no negative value

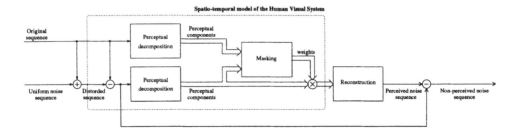

Fig. 3. *Generation of the non-perceived noise sequence npn(x,y,t).*

In other words, if we define the perceptual activity measure, $MPAM(x, y, t)$, of a block centered around (x, y) of size (N_x, N_y) at time t as the mean luminance value of the corresponding block in the non-perceived noise sequence :

$$MPAM(x, y, t; N_x, N_y) \triangleq \frac{1}{N_x N_y} \left(\sum_{a=x-N_x/2}^{x+N_x/2} \sum_{b=y-N_y/2}^{y+N_y/2} npn(a, b, t) \right) ,$$

zones where $MPAM(x, y, t)$ is high correspond to equivalent zones in the original sequence which are relevant for human perception, and conversely [11].

This activity metric gives us correlated results with human judgements. However, the temporal complexity of such a metric is too high to be used within multimedia applications. This is mainly due to the computation of the two perceptual decompositions. Furthermore, as we need to reconstruct the data (reconstruction block), we must use the perfect reconstruction bank, to alleviate scalloping effect, which features more subbands and is only an approximation of the Gabor filter bank [5].

3.2 Perceptual Visibility Predictor (PVP)

Our Perceptual Visibility Predictor (PVP) is an optimized tool, based on insights about visual masking, that predicts the perceptual relevance of sequence areas.

A common model of masking is a non-linear transducer in which two regions can be identified. Let us denote C_{T0} the detection threshold of the signal in the absence of masker, C_T the detection threshold of the signal in the presence of the masker and C_M the contrast of the masker.

The detection threshold first remains constant at low values of C_M ($C_T = C_{T0}$) and, as C_M increases, increases as a power of the contrast masker. This function is linear in a log-log graph and its slope is denoted ϵ (see Fig. 4).

The actual detection threshold, C_T is then computed as :

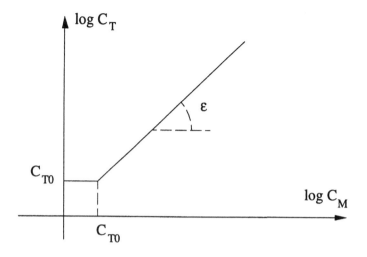

Fig. 4. *Model of the masking phenomenon.*

$$C_T = \begin{cases} C_{T0}, & \text{if } C_M < C_{T0} \\ C_{T0} \cdot [\frac{C_M}{C_{T0}}]^\epsilon, & \text{otherwise.} \end{cases} \tag{1}$$

Let us consider now the block diagram presented in Fig. 5. The video sequence is first decomposed into perceptual channels by a three-dimensional filter bank that emulates the detection mechanisms of the cortex. The next block models pattern sensitivity, i.e. appearance of the stimuli as a function of their frequency. In our model, pattern sensitivity computes the *detection threshold for the coding noise* due to the scene[2]. This is equivalent to computing the local perceptual activity of the scene, on a channel basis.

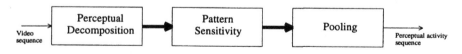

Fig. 5. *Perceptual Visibility Predictor (PVP) block diagram. The thick arrows represent a set of perceptual components while the thin ones represent video sequences.*

The following block, denoted *pooling*, simulates higher order integration by subsequent areas of the cortex. It gathers the channels measurements together

---

[2] the scene can be considered as a masker with respect to coding noise, i.e. part of the noise is masked by the scene.

to yield a single figure. This measurement is termed the *visibility predictor* as it predicts the perceptual relevance of a region. Let $C_T(a, b, t, c)$ be the computed detection threshold at position (a, b), time t and in channel c. The Perceptual Visibility Predictor (PVP) for a region of spatial dimension N_x and N_y, centered around (x, y) at time t is computed as :

$$PVP(x, y, t; N_x, N_y) = \left(\frac{1}{N} \sum_{c=1}^{N} \left(\frac{1}{N_x N_y} \sum_{a=x-N_x/2}^{x+N_x/2} \sum_{b=y-N_y/2}^{y+N_y/2} |C_T[a, b, t, c]| \right)^{\beta} \right)^{\frac{1}{\beta}} ,$$

where, $N = 34$ is the number of channels and β has a value of 4 [3].

Figure 7 shows the PVP map $(N_x = N_y = 16)$ corresponding to the 10th frame out of 64 of the Basket-Ball sequence (see Fig. 6). Due to the spatio-temporal masking phenomenon, in a given block, the higher the PVP value, the lower the annoying impact of a coding or transmission error.

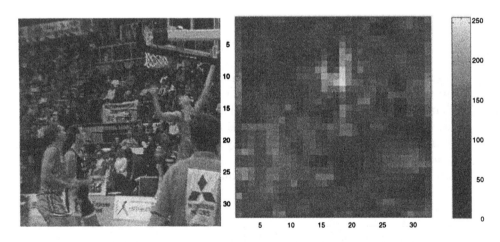

Fig. 6. Frame from Basketball sequence. **Fig. 7.** Perceptual Visibility Predictor (PVP) map computed over (16×16)-size blocks.

Both metrics give results correlated with human judgements. However, unlike the preceding one, the PVP metric is optimized for use in multimedia applications, since the temporal complexity is now sufficiently low compared to the complexity of coding algorithms. Compared to the MPAM metric, it requires a single perceptual decomposition with a fewer number of channels and does not mandate any additional synthetic sequence, nor any reconstruction process.

4 Optimization of Video Services based on MPEG-2 Encoding

This section motivates the use of both perceptual video quality and activity metrics in order to optimize the performance of video services in terms of the end-user quality perception. The video encoding scheme considered in this paper is MPEG-2 (acronym standing for "Motion Picture Experts Group"). MPEG-2 is the ISO/IEC standard [12] for full-motion, high-quality video and has been designed to satisfy a large variety of video applications.

We first describe the essential features of the MPEG-2 compression standard. We then analyze the impact of data loss on the reconstructed MPEG-2 sequence. Finally, based on this knowledge, we propose some applications of our perceptual quality and activity metrics on MPEG-2 based video services.

4.1 The MPEG-2 Video and System Standards

The structure of an MPEG-2 video stream is illustrated in Fig. 8. The stream consists of a sequence which is composed of several frames. Each frame is composed of slices which are, by definition, a serie of macroblocks. Each macroblock (16 × 16 pixels) contains 4 blocks (8 × 8 pixels) of luminance and 2,4 or 8 blocks of chrominance depending on the chroma format used.

The MPEG-2 video syntax defines three different types of frames :

- Intra (or I) pictures, are coded using information only found in the picture itself. I-frames provide potential random access points into the compressed video data. I-frames only use Discrete Cosine Transform (DCT) on 8 × 8 blocks. The DCT coefficients are then quantized. I-frames provide moderate compression ratios.
- Predicted (or P) pictures are coded with respect to the nearest previous I or P-frame. This technique is called forward prediction. Like I-frames, P-frames serve as a prediction reference for B-frames and future P-frames. However, P-frames use motion compensation to get a higher compression ratio then is possible for I-frames.
- Bidirectional (or B) pictures use both a past and future picture as a reference. This technique is called bidirectional prediction. B-frames provide the highest compression. Bidirectional prediction also reduces the effects of noise by averaging two pictures.

The standard does not specify how I-, P- and B-frames have to be mixed together. This can be set by the user. The use of these three frame types allows MPEG-2 to be robust (I-frames provide error propagation reset points) and efficient (B- and P-frames allow a good overall compression ratio).

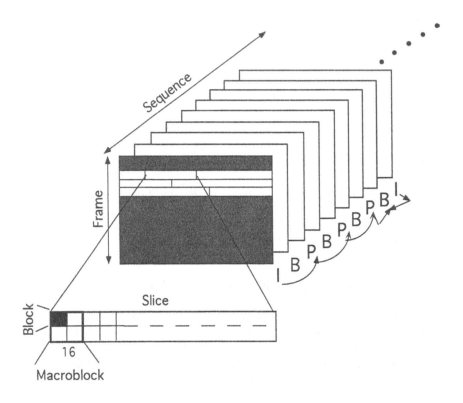

Fig. 8. *MPEG-2 Video Structure.*

MPEG-2 coding of video naturally produces a varying bit rate since not all video frames have the same entropy [13]. However, the Test Model v5 (TM5) proposes a rate control algorithm producing a constant bit rate (CBR) MPEG-2 stream [10].

The MPEG-2 system document [14] specifies two systems. The first multiplexes together the video, audio and data of single program to be transmitted in a relatively error-free environment into the program stream. Another system, the transport stream, can be used for broadcast, VOD and cable TV.

The transport stream defines a packetized protocol for multiplexing multiple MPEG-2 compressed programs into a packetized fixed-length (188 bytes) format (Transport Stream or TS packets) for transmission on digital networks. It also includes some sophisticated timing information, such as jitter correction, etc. The additional timing information and small fixed-sized packets allow a whole range of new application for MPEG-2. The 188-byte TS packets can map very well into 48-byte ATM cell payloads, allowing MPEG-2 to be used in switched video architecture.

4.2 Perceptual Impact of Cell Losses

In an MPEG-2 video stream, the reduction of quality due to data loss strongly depends on the type of the lost information. Losses in syntactic data, such as headers and system information, affect the quality differently than losses of semantic data such as pure video information. Furthermore, the quality reduction depends also on the location of the lost semantic data due to the predictive structure of an MPEG-2 video coded stream.

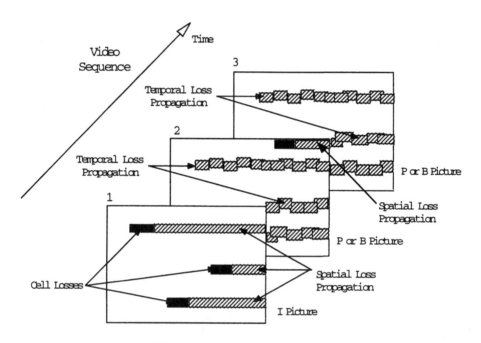

Fig. 9. *Data Loss Propagation.*

Let us consider Fig. 9, showing how network losses map onto visual information losses in different types of MPEG-2 pictures. Indeed, data loss spreads within a single picture up to the next resynchronization point (e.g. slice headers) mainly due to the use of variable length coding, run length coding and differential coding. This is referred to as spatial propagation and may damage any type of picture. When loss occurs in a reference picture (I- or P-frame), it will damage the quality until the next intra-coded picture is received. This is known as temporal propagation and is due to inter-frame predictions.

The impact the loss of syntactic data may have is in general more important and more difficult to recover than the loss of semantic information. This data loss may induce frame loss in the decoded sequence. Indeed, when a frame header (a few syntactic bytes before each frame in the bitstream) is lost, the entire corresponding frame is skipped because the decoder is not able to detect the beginning of the frame. If the skipped frame corresponds to a predictive picture (I- or P-frame), it may strongly reduce the perceptual quality.

The problem is actually that when a header is lost, in general, the whole information it carries is skipped. Some headers are thus more crucial than others. For instance, sequence headers, predictive (I or P) picture headers, PES headers and slice headers in intra-coded pictures can be considered essential in comparison to slice headers in B-frames.

Error concealment algorithms have already shown that it is possible to reduce the impact of data loss on the visual information [15, 16, 17]. These error concealment algorithms include, for example, spatial interpolation, temporal interpolation and early resynchronization techniques. Early resynchronization decoding techniques limit the spatial propagation of errors by decoding some semantic information that is normally discarded from the damaged MPEG-2 video streams. In other words, it helps the decoder to resynchronize quickly (at least before the following header). This method is based on the identification of allowed codewords as proposed in [16] and, unfortunately, only works with intra-coded frames.

4.3 Applications of Both Metrics

In this section, we describe three different application fields of both the perceptual video quality (i.e. MPQM) and activity (i.e. PVP) metrics. We actually motivate the use of psychophysics in order to improve MPEG-2 based video services on both the encoding and transmission levels.

Perceptual Quality Metric as a Performance Tool for ATM Adaptation of MPEG-2 based Multimedia Applications

In [18, 19] the authors studied the impact of data loss on MPEG-2 video coded streams transmitted over an ATM network on top of AAL5 and also on new versions of both network and ATM adaptation layers. The impact has been measured using the perceptual quality metric presented in section 2. Figure 10 shows the comparison of the behaviour of both AAL5 and the new proposed AAL (AAL-RT) when the MPEG-2 stream is subject to data loss (i.e. cell losses).

The MPEG-2 stream consists of a ski sequence of 1000 frames (720×576) encoded at 4MBits/s as interlaced video, with a structure of 12 images per GOP and 2 B-pictures between every reference picture, and a single slice per line.

Figure 11 presents the MPQM quality assessment of MPEG-2 video for the same ski sequence as a function of the encoding bit rate. The behaviour of the curves presented in Fig. 10 and Fig. 11 may lead to an interesting study. Indeed, cell losses do not have the same annoying impact on sequences coded at 3 Mbits/s as opposed to 7 Mbits/s. Furthermore, the perceptual coding quality

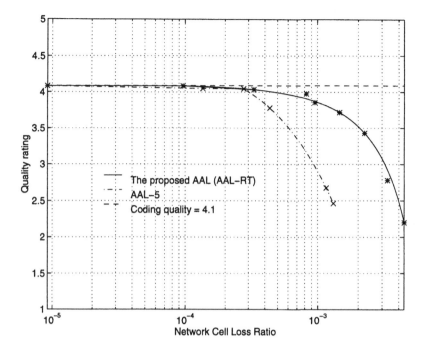

Fig. 10. *Quality rating versus network cell loss ratio for the new proposed AAL and for AAL5.*

is very different (from less than 3 to almost 4.5). Consequently, one can find the optimum working bit rate and cell loss ratio for maximizing the end-user perception of the video quality.

The perceptual quality metric also allows the comparison of the efficiency of error concealment techniques. For instance, we show in Fig. 12 the difference in perceptual quality when early resynchronization algorithms are used or not with the new proposed AAL.

Bit Allocation for CBR MPEG-2 Encoding based on Psychophysics

Bit allocation based on psychophysics (or perceptual bit allocation, PBA) means to allocate more bits for perceptually relevant macroblocks while assigning fewer number of bits for macroblocks in which most of the error would be masked by human vision. This provides a good example to test the efficiency of our PVP metric described in section 3.2.

As stated before, MPEG-2 encoding can provide both CBR and VBR video streams. The CBR MPEG-2 encoding scheme now receives a lot of attention since some video services, such as the Video on Demand service, are based on a

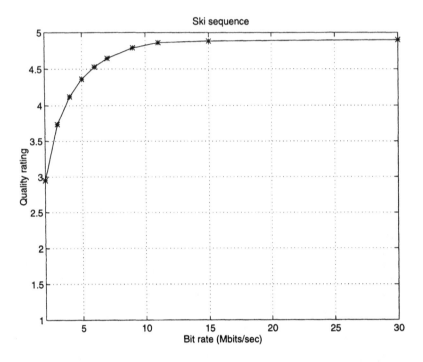

Fig. 11. *MPQM quality assessment for the ski sequence as a function of the bit rate.*

CBR transmission. The CBR encoding must be guaranteed by a rate controller usually based on the TM5 rate control algorithm [10].

The TM5 rate control algorithm is divided into three steps, namely target bit allocation, rate control via buffer monitoring and adaptive quantization. The third step computes a quantization value (i.e. *mquant*) by modulating the macroblock quantization step size by a local activity measure estimated by means of a block-based variance measure. As stated in [11], the variance measure is poorly correlated with human perception and cannot be used to reliably predict the activity in blocks. Therefore, we replaced the bit allocation as proposed in the MPEG-2 TM5 with our new perceptual bit allocation using the PVP metric. This solution is still fully compatible with the MPEG-2 standard.

Simulations have been performed on 64 frames of the Basket-Ball sequence. This sequence has been compressed at 15 different rates between 1 and 15 Mbits/s.

We present results in terms of quality measurements of the resulting compressed streams versus the bit rate. Figure 13 shows performance comparison of the bit allocation scheme using PVP and the bit allocation scheme as proposed in MPEG-2 TM5. It can be seen that for a given quality, the resulting bit stream

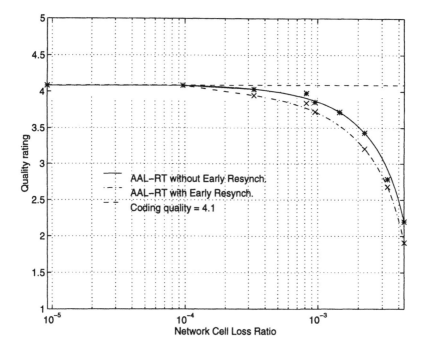

Fig. 12. *Quality rating versus network cell loss ratio for the new proposed AAL with and without the using early resynchronization mechanisms.*

uses less bandwidth with our scheme (almost 1 Mbits/s for usual encoding bit rates). On the contrary, for a given bandwidth, the scheme results in a video stream of higher perceptual quality.

Perceptual Syntactic Information MPEG-2 Coding (PSIC)

As stated in section 4.2, data loss spreads within a single picture up to the next resynchronization point (i.e. next slice header). As shown in Fig. 8, the MPEG-2 standard allows to build slices with a variable number of macroblocks. However, the greater the number of slices (i.e. number of slice headers), the bigger the overhead. Therefore, the perceptual syntactic information coding (PSIC) tries, for a given overhead, to structure syntactic information such as slice headers in order to minimize the annoying impact of data loss on the end-user perception of quality. For instance, slices are built according to the PVP map. A new slice header is inserted as soon as the sum of the macroblock-based PVP values reaches a given threshold. This threshold is a function of the frame type (I-, P- or B- frame), the expected cell loss ratio (CLR), the expected overhead and the negociated perceptual quality.

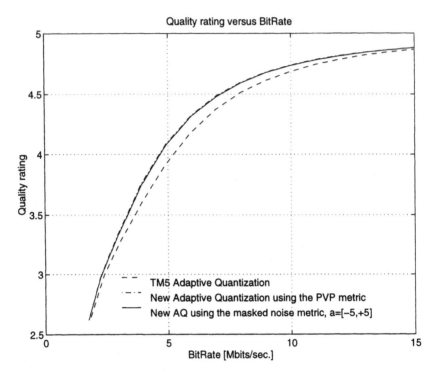

Fig. 13. *Perceptual video quality versus MPEG-2 encoding bit rate.*

5 Conclusion

In this paper, we have considered the optimization of MPEG-2 based video services by means of a perceptual video quality metric (i.e. MPQM) and a new video activity metric (i.e. PVP) based on an advanced model of the human vision.

We have first described a spatio-temporal model of the human visual system from which a perceptual video quality metric has been derived. We have then presented two perceptual local video activity metrics, namely the MPAM and PVP metrics. We have finally studied how these quality and activity metrics could be used to optimize video services based on MPEG-2 encoding by means of three different examples.

Some further research will be carried out to design some other mechanisms using psychophysics. We are currently working on several new perceptual quality control, rate control and reliability control algorithms for both CBR and VBR transmissions.

6 Acknowledgments

The authors would like to thank Pascal Frossard for his helpful collaboration in performing the simulations.

References

1. J. A. Saghri, P. S. Cheatham and A. Habibi, "Image Quality measure based on a Human Visual System Model", *in Optical Engineering*, vol. 28, pp. 813–818, 1989.
2. S. Comes and B. Macq, "Human Visual Quality Criterion", *SPIE Visual Communications and Image Processing*, vol. 1360, pp. 2–7, 1990.
3. Christian J. van den Branden Lambrecht and Olivier Verscheure, "Perceptual Quality Measure using a Spatio-Temporal Model of the Human Visual System", *in Proceedings of the SPIE*, vol. 2668, pp. 450–461, San Jose, CA, January 28 - February 2 1996, available on http://tcomwww.epfl.ch/~verscheu.
4. Serge Comes, *Les traitements perceptifs d'images numerisées*, PhD thesis, Université Catholique de Louvain, 1995.
5. Christian J. van den Branden Lambrecht, "A Working Spatio-Temporal Model of the Human Visual System for Image Restoration and Quality Assessment Applications", *in Proceedings of the International Conference on Acoustics, Speech, and Signal Processing*, pp. 2293–2296, Atlanta, GA, May 7-10 1996, available on http://ltswww.epfl.ch/pub_files/vdb/.
6. Christian J. van den Branden Lambrecht, *Perceptual Models and Architectures for Video Coding Applications*, PhD thesis, Swiss Federal Institute of Technology, CH 1015 Lausanne, Switzerland, 1996, available on http://ltswww.epfl.ch/pub_files/vdb/.
7. M. Ardito and M. Barbero and M. Stroppiana and M. Visca, "Compression and Quality", *in Proceedings of the International Workshop on HDTV 94*, Brisbane, Australia, October 1994.
8. Pär Lindh and Christian J. van den Branden Lambrecht, "Efficient Spatio-Temporal Decomposition for Perceptual Processing of Video Sequences", *in Proceedings of the International Conference on Image Processing*, Lausanne, Switzerland, September 16-19 1996, available on http://ltswww.epfl.ch/pub_files/vdb/.
9. David J. Heeger and Patrick C. Teo, "A Model of Perceptual Image Fidelity", *in Proceedings of the International Conference on Image Processing*, pp. 343–345, Washington, DC, October 23-26 1995.
10. Chadd Fogg, "mpeg2encode/mpeg2decode version 1.1. available via anonymous ftp at ftp.netcom.com", *in MPEG Software Simulation Group*, June 1994.
11. Olivier Verscheure, Andrea Basso, Mounir El-Maliki and Jean-Pierre Hubaux, "Perceptual Bit Allocation for MPEG-2 Video Coding", *in Proceedings of the International Conference on Image Processing*, Lausanne, Switzerland, September 16-19 1996, available on http://tcomwww.epfl.ch/~verscheu.
12. ISO/IEC JTC 1, *Information Technology - Generic Coding of Moving Pictures and Associated Audio Information - Part 2: Video*, ISO/IEC JTC 1, 1994.
13. D. Reininger, et al., "Variable bitrate mpeg video : Characteristics, modeling and multiplexing", *in Proceedings of the ITC-14*, pp. 295–306. Elsevier, 1994.

14. ISO/IEC JTC 1, *Information Technology - Generic Coding of Moving Pictures and Associated Audio Information - Part 1: Systems Specification*, ISO/IEC JTC 1, 1994.

15. Carlos Lopez Fernandez, Andrea Basso and Jean-Pierre Hubaux, "Error Concealment and Early Resynchronization Techniques for MPEG-2 Video Streams Damaged by Transmission over ATM Networks", *in SPIE*, San Jose, USA, February 1996.

16. S. Lee, J. Youn and S. Jang, "Transmission error detection, resynchronization, and error concealment for mpeg-2 video decoder", *in SPIE*, 1993.

17. S. Aign and K. Fazel, "Temporal and spatial error concealment techniques for hierarchical mpeg-2 video coded", *in ICC'95*, 1995.

18. Xavier Garcia Adanez, Olivier Verscheure and Jean-Pierre Hubaux, "New Network and ATM Adaptation Layers for Real-Time Multimedia Applications: A Performance Study Based on Psychophysics", *in COST-237 workshop*, Barcelona, Spain, November 1996, available on http://tcomwww.epfl.ch/~verscheu.

19. Olivier Verscheure and Xavier Garcia Adanez, "Perceptual Quality Metric as a Performance Tool for ATM Adaptation of MPEG-2 based Multimedia Applications", *in EUNICE Summer School on Telecommunications Services*, Lausanne, Switzerland, September 1996, available on http://tcomwww.epfl.ch/~verscheu.

Index of Authors

Á. Almeida, 130
A. Azcorra, 137
D. Bourges Waldegg, 232
M. C. Chan, 56
G. Coulson, 75
J. S. Crawford, 23
T. de Miguel, 137
M. Degermark, 169
F. Eryurtlu, 183
P. Freiburghaus, 1
X. Garcia Adanez, 216
R. Hirn, 154
M. Hofmann, 41
J.-F. Huard, 56
J.-P. Hubaux, 216, 249
G. Huecas, 137
R. J. Huis in't Veld, 201
D. Hutchison, 75
D. Koller, 1
A. M. Kondoz, 183
A.-N. Ladhani, 201
N. Lagha, 232
D. Larrabeiti, 137

A. A. Lazar, 56
J.-P. Le Narzul, 232
H. Leopold, 154
K.-S. Lim, 56
F. Moelaert El-Hadidy, 201
S. Pavon, 137
M. Petit, 137
T. Pfeifer, 104
S. Pink, 169
B. Plattner, 1
R. Popescu-Zeletin, 104
J. Quemada, 137
T. Robles, 137
A. H. Sadka, 183
J. Salvachua, 137
B. van der Waaij, 201
J. P. C. Verhoosel, 201
O. Verscheure, 216, 249
D. G. Waddington, 75
A. G. Waters, 23
I. A. Widya, 201
E. Wilde, 1

Springer
and the
environment

At Springer we firmly believe that an international science publisher has a special obligation to the environment, and our corporate policies consistently reflect this conviction.

We also expect our business partners – paper mills, printers, packaging manufacturers, etc. – to commit themselves to using materials and production processes that do not harm the environment. The paper in this book is made from low- or no-chlorine pulp and is acid free, in conformance with international standards for paper permanency.

Lecture Notes in Computer Science

For information about Vols. 1–1107

please contact your bookseller or Springer-Verlag

Vol. 1108: A. Díaz de Ilarraza Sánchez, I. Fernández de Castro (Eds.), Computer Aided Learning and Instruction in Science and Engineering. Proceedings, 1996. XIV, 480 pages. 1996.

Vol. 1109: N. Koblitz (Ed.), Advances in Cryptology – Crypto '96. Proceedings, 1996. XII, 417 pages. 1996.

Vol. 1110: O. Danvy, R. Glück, P. Thiemann (Eds.), Partial Evaluation. Proceedings, 1996. XII, 514 pages. 1996.

Vol. 1111: J.J. Alferes, L. Moniz Pereira, Reasoning with Logic Programming. XXI, 326 pages. 1996. (Subseries LNAI).

Vol. 1112: C. von der Malsburg, W. von Seelen, J.C. Vorbrüggen, B. Sendhoff (Eds.), Artificial Neural Networks – ICANN 96. Proceedings, 1996. XXV, 922 pages. 1996.

Vol. 1113: W. Penczek, A. Szałas (Eds.), Mathematical Foundations of Computer Science 1996. Proceedings, 1996. X, 592 pages. 1996.

Vol. 1114: N. Foo, R. Goebel (Eds.), PRICAI'96: Topics in Artificial Intelligence. Proceedings, 1996. XXI, 658 pages. 1996. (Subseries LNAI).

Vol. 1115: P.W. Eklund, G. Ellis, G. Mann (Eds.), Conceptual Structures: Knowledge Representation as Interlingua. Proceedings, 1996. XIII, 321 pages. 1996. (Subseries LNAI).

Vol. 1116: J. Hall (Ed.), Management of Telecommunication Systems and Services. XXI, 229 pages. 1996.

Vol. 1117: A. Ferreira, J. Rolim, Y. Saad, T. Yang (Eds.), Parallel Algorithms for Irregularly Structured Problems. Proceedings, 1996. IX, 358 pages. 1996.

Vol. 1118: E.C. Freuder (Ed.), Principles and Practice of Constraint Programming — CP 96. Proceedings, 1996. XIX, 574 pages. 1996.

Vol. 1119: U. Montanari, V. Sassone (Eds.), CONCUR '96: Concurrency Theory. Proceedings, 1996. XII, 751 pages. 1996.

Vol. 1120: M. Deza. R. Euler, I. Manoussakis (Eds.), Combinatorics and Computer Science. Proceedings, 1995. IX, 415 pages. 1996.

Vol. 1121: P. Perner, P. Wang, A. Rosenfeld (Eds.), Advances in Structural and Syntactical Pattern Recognition. Proceedings, 1996. X, 393 pages. 1996.

Vol. 1122: H. Cohen (Ed.), Algorithmic Number Theory. Proceedings, 1996. IX, 405 pages. 1996.

Vol. 1123: L. Bougé, P. Fraigniaud, A. Mignotte, Y. Robert (Eds.), Euro-Par'96. Parallel Processing. Proceedings, 1996, Vol. I. XXXIII, 842 pages. 1996.

Vol. 1124: L. Bougé, P. Fraigniaud, A. Mignotte, Y. Robert (Eds.), Euro-Par'96. Parallel Processing. Proceedings, 1996, Vol. II. XXXIII, 926 pages. 1996.

Vol. 1125: J. von Wright, J. Grundy, J. Harrison (Eds.), Theorem Proving in Higher Order Logics. Proceedings, 1996. VIII, 447 pages. 1996.

Vol. 1126: J.J. Alferes, L. Moniz Pereira, E. Orlowska (Eds.), Logics in Artificial Intelligence. Proceedings, 1996. IX, 417 pages. 1996. (Subseries LNAI).

Vol. 1127: L. Böszörményi (Ed.), Parallel Computation. Proceedings, 1996. XI, 235 pages. 1996.

Vol. 1128: J. Calmet, C. Limongelli (Eds.), Design and Implementation of Symbolic Computation Systems. Proceedings, 1996. IX, 356 pages. 1996.

Vol. 1129: J. Launchbury, E. Meijer, T. Sheard (Eds.), Advanced Functional Programming. Proceedings, 1996. VII, 238 pages. 1996.

Vol. 1130: M. Haveraaen, O. Owe, O.-J. Dahl (Eds.), Recent Trends in Data Type Specification. Proceedings, 1995. VIII, 551 pages. 1996.

Vol. 1131: K.H. Höhne, R. Kikinis (Eds.), Visualization in Biomedical Computing. Proceedings, 1996. XII, 610 pages. 1996.

Vol. 1132: G.-R. Perrin, A. Darte (Eds.), The Data Parallel Programming Model. XV, 284 pages. 1996.

Vol. 1133: J.-Y. Chouinard, P. Fortier, T.A. Gulliver (Eds.), Information Theory and Applications II. Proceedings, 1995. XII, 309 pages. 1996.

Vol. 1134: R. Wagner, H. Thoma (Eds.), Database and Expert Systems Applications. Proceedings, 1996. XV, 921 pages. 1996.

Vol. 1135: B. Jonsson, J. Parrow (Eds.), Formal Techniques in Real-Time and Fault-Tolerant Systems. Proceedings, 1996. X, 479 pages. 1996.

Vol. 1136: J. Diaz, M. Serna (Eds.), Algorithms – ESA '96. Proceedings, 1996. XII, 566 pages. 1996.

Vol. 1137: G. Görz, S. Hölldobler (Eds.), KI-96: Advances in Artificial Intelligence. Proceedings, 1996. XI, 387 pages. 1996. (Subseries LNAI).

Vol. 1138: J. Calmet, J.A. Campbell, J. Pfalzgraf (Eds.), Artificial Intelligence and Symbolic Mathematical Computation. Proceedings, 1996. VIII, 381 pages. 1996.

Vol. 1139: M. Hanus, M. Rogriguez-Artalejo (Eds.), Algebraic and Logic Programming. Proceedings, 1996. VIII, 345 pages. 1996.

Vol. 1140: H. Kuchen, S. Doaitse Swierstra (Eds.), Programming Languages: Implementations, Logics, and Programs. Proceedings, 1996. XI, 479 pages. 1996.

Vol. 1141: H.-M. Voigt, W. Ebeling, I. Rechenberg, H.-P. Schwefel (Eds.), Parallel Problem Solving from Nature – PPSN IV. Proceedings, 1996. XVII, 1.050 pages. 1996.

Vol. 1142: R.W. Hartenstein, M. Glesner (Eds.), Field-Programmable Logic. Proceedings, 1996. X, 432 pages. 1996.

Vol. 1143: T.C. Fogarty (Ed.), Evolutionary Computing. Proceedings, 1996. VIII, 305 pages. 1996.

Vol. 1144: J. Ponce, A. Zisserman, M. Hebert (Eds.), Object Representation in Computer Vision. Proceedings, 1996. VIII, 403 pages. 1996.

Vol. 1145: R. Cousot, D.A. Schmidt (Eds.), Static Analysis. Proceedings, 1996. IX, 389 pages. 1996.

Vol. 1146: E. Bertino, H. Kurth, G. Martella, E. Montolivo (Eds.), Computer Security – ESORICS 96. Proceedings, 1996. X, 365 pages. 1996.

Vol. 1147: L. Miclet, C. de la Higuera (Eds.), Grammatical Inference: Learning Syntax from Sentences. Proceedings, 1996. VIII, 327 pages. 1996. (Subseries LNAI).

Vol. 1148: M.C. Lin, D. Manocha (Eds.), Applied Computational Geometry. Proceedings, 1996. VIII, 223 pages. 1996.

Vol. 1149: C. Montangero (Ed.), Software Process Technology. Proceedings, 1996. IX, 291 pages. 1996.

Vol. 1150: A. Hlawiczka, J.G. Silva, L. Simoncini (Eds.), Dependable Computing – EDCC-2. Proceedings, 1996. XVI, 440 pages. 1996.

Vol. 1151: Ö. Babaoğlu, K. Marzullo (Eds.), Distributed Algorithms. Proceedings, 1996. VIII, 381 pages. 1996.

Vol. 1152: T. Furuhashi, Y. Uchikawa (Eds.), Fuzzy Logic, Neural Networks, and Evolutionary Computation. Proceedings, 1995. VIII, 243 pages. 1996. (Subseries LNAI).

Vol. 1153: E. Burke, P. Ross (Eds.), Practice and Theory of Automated Timetabling. Proceedings, 1995. XIII, 381 pages. 1996.

Vol. 1154: D. Pedreschi, C. Zaniolo (Eds.), Logic in Databases. Proceedings, 1996. X, 497 pages. 1996.

Vol. 1155: J. Roberts, U. Mocci, J. Virtamo (Eds.), Broadbank Network Teletraffic. XXII, 584 pages. 1996.

Vol. 1156: A. Bode, J. Dongarra, T. Ludwig, V. Sunderam (Eds.), Parallel Virtual Machine – EuroPVM '96. Proceedings, 1996. XIV, 362 pages. 1996.

Vol. 1157: B. Thalheim (Ed.), Conceptual Modeling – ER '96. Proceedings, 1996. XII, 489 pages. 1996.

Vol. 1158: S. Berardi, M. Coppo (Eds.), Types for Proofs and Programs. Proceedings, 1995. X, 296 pages. 1996.

Vol. 1159: D.L. Borges, C.A.A. Kaestner (Eds.), Advances in Artificial Intelligence. Proceedings, 1996. XI, 243 pages. (Subseries LNAI).

Vol. 1160: S. Arikawa, A.K. Sharma (Eds.), Algorithmic Learning Theory. Proceedings, 1996. XVII, 337 pages. 1996. (Subseries LNAI).

Vol. 1161: O. Spaniol, C. Linnhoff-Popien, B. Meyer (Eds.), Trends in Distributed Systems. Proceedings, 1996. VIII, 289 pages. 1996.

Vol. 1162: D.G. Feitelson, L. Rudolph (Eds.), Job Scheduling Strategies for Parallel Processing. Proceedings, 1996. VIII, 291 pages. 1996.

Vol. 1163: K. Kim, T. Matsumoto (Eds.), Advances in Cryptology – ASIACRYPT '96. Proceedings, 1996. XII, 395 pages. 1996.

Vol. 1164: K. Berquist, A. Berquist (Eds.), Managing Information Highways. XIV, 417 pages. 1996.

Vol. 1165: J.-R. Abrial, E. Börger, H. Langmaack (Eds.), Formal Methods for Industrial Applications. VIII, 511 pages. 1996.

Vol. 1166: M. Srivas, A. Camilleri (Eds.), Formal Methods in Computer-Aided Design. Proceedings, 1996. IX, 470 pages. 1996.

Vol. 1167: I. Sommerville (Ed.), Software Configuration Management. VII, 291 pages. 1996.

Vol. 1168: I. Smith, B. Faltings (Eds.), Advances in Case-Based Reasoning. Proceedings, 1996. IX, 531 pages. 1996. (Subseries LNAI).

Vol. 1169: M. Broy, S. Merz, K. Spies (Eds.), Formal Systems Specification. XXIII, 541 pages. 1996.

Vol. 1170: M. Nagl (Ed.), Building Tightly Integrated Software Development Environments: The IPSEN Approach. IX, 709 pages. 1996.

Vol. 1171: A. Franz, Automatic Ambiguity Resolution in Natural Language Processing. XIX, 155 pages. 1996. (Subseries LNAI).

Vol. 1172: J. Pieprzyk, J. Seberry (Eds.), Information Security and Privacy. Proceedings, 1996. IX, 333 pages. 1996.

Vol. 1173: W. Rucklidge, Efficient Visual Recognition Using the Hausdorff Distance. XIII, 178 pages. 1996.

Vol. 1174: R. Anderson (Ed.), Information Hiding. Proceedings, 1996. VIII, 351 pages. 1996.

Vol. 1175: K.G. Jeffery, J. Král, M. Bartošek (Eds.), SOFSEM'96: Theory and Practice of Informatics. Proceedings, 1996. XII, 491 pages. 1996.

Vol. 1176: S. Miguet, A. Montanvert, S. Ubéda (Eds.), Discrete Geometry for Computer Imagery. Proceedings, 1996. XI, 349 pages. 1996.

Vol. 1177: J.P. Müller, The Design of Intelligent Agents. XV, 227 pages. 1996. (Subseries LNAI).

Vol. 1178: T. Asano, Y. Igarashi, H. Nagamochi, S. Miyano, S. Suri (Eds.), Algorithms and Computation. Proceedings, 1996. X, 448 pages. 1996.

Vol. 1179: J. Jaffar, R.H.C. Yap (Eds.), Concurrency and Parallelism, Programming, Networking, and Security. Proceedings, 1996. XIII, 394 pages. 1996.

Vol. 1180: V. Chandru, V. Vinay (Eds.), Foundations of Software Technology and Theoretical Computer Science. Proceedings, 1996. XI, 387 pages. 1996.

Vol. 1181: D. Bjørner, M. Broy, I.V. Pottosin (Eds.), Perspectives of System Informatics. Proceedings, 1996. XVII, 447 pages. 1996.

Vol. 1182: W. Hasan, Optimization of SQL Queries for Parallel Machines. XVIII, 133 pages. 1996.

Vol. 1183: A. Wierse, G.G. Grinstein, U. Lang (Eds.), Database Issues for Data Visualization. Proceedings, 1995. XIV, 219 pages. 1996.

Vol. 1184: J. Waśniewski, J. Dongarra, K. Madsen, D. Olesen (Eds.), Applied Parallel Computing. Proceedings, 1996. XIII, 722 pages. 1996.

Vol. 1185: G. Ventre, J. Domingo-Pascual, A. Danthine (Eds.), Multimedia Telecommunications and Applications. Proceedings, 1996. XII, 267 pages. 1996.

Vol. 1186: F. Afrati, P. Kolaitis (Eds.), Database Theory – ICDT'97. Proceedings, 1997. XIII, 477 pages. 1997.